Advances in Intelligent Systems and Computing

Volume 1357

AF831523

The series "Advances in Intelligent Systems and Computing" contains publications on theory, applications, and design methods of Intelligent Systems and Intelligent Computing. Virtually all disciplines such as engineering, natural sciences, computer and information science, ICT, economics, business, e-commerce, environment, healthcare, life science are covered. The list of topics spans all the areas of modern intelligent systems and computing such as: computational intelligence, soft computing including neural networks, fuzzy systems, evolutionary computing and the fusion of these paradigms, social intelligence, ambient intelligence, computational neuroscience, artificial life, virtual worlds and society, cognitive science and systems, Perception and Vision, DNA and immune based systems, self-organizing and adaptive systems, e-Learning and teaching, human-centered and human-centric computing, recommender systems, intelligent control, robotics and mechatronics including human-machine teaming, knowledge-based paradigms, learning paradigms, machine ethics, intelligent data analysis, knowledge management, intelligent agents, intelligent decision making and support, intelligent network security, trust management, interactive entertainment, Web intelligence and multimedia.

The publications within "Advances in Intelligent Systems and Computing" are primarily proceedings of important conferences, symposia and congresses. They cover significant recent developments in the field, both of a foundational and applicable character. An important characteristic feature of the series is the short publication time and world-wide distribution. This permits a rapid and broad dissemination of research results.

Indexed by DBLP, INSPEC, WTI Frankfurt eG, zbMATH, Japanese Science and Technology Agency (JST).

All books published in the series are submitted for consideration in Web of Science.

More information about this series at http://www.springer.com/series/11156

Katsutoshi Yada · Daisuke Katagami ·
Yasufumi Takama · Takayuki Ito ·
Akinori Abe · Eri Sato-Shimokawara ·
Junichiro Mori · Naohiro Matsumura ·
Hisashi Kashima
Editors

Advances in Artificial Intelligence

Selected Papers from the Annual Conference of Japanese Society of Artificial Intelligence (JSAI 2020)

Springer

Editors
Katsutoshi Yada
Faculty of Business and Commerce
Kansai University
Suita, Osaka, Japan

Daisuke Katagami
Department of Applied Computer Science
Tokyo Polytechnic University
Atsugi, Kanagawa, Japan

Yasufumi Takama
Graduate School of System Design
Tokyo Metropolitan University
Hino, Tokyo, Japan

Takayuki Ito
Department of Social Informatics
Kyoto University
Kyoto, Japan

Akinori Abe
Division of Behavioral Science,
Faculty of Letters
Chiba University
Chiba, Chiba, Japan

Eri Sato-Shimokawara
Department of Computer Science,
Graduate School of System Design
Tokyo Metropolitan University
Hino, Tokyo, Japan

Junichiro Mori
Mathematics and Informatics Center
The University of Tokyo
Tokyo, Japan

Naohiro Matsumura
Graduate School of Economics
Osaka University
Toyonaka, Osaka, Japan

Hisashi Kashima
Department of Intelligence Science
and Technology
Kyoto University
Kyoto, Japan

ISSN 2194-5357 ISSN 2194-5365 (electronic)
Advances in Intelligent Systems and Computing
ISBN 978-3-030-73112-0 ISBN 978-3-030-73113-7 (eBook)
https://doi.org/10.1007/978-3-030-73113-7

This Springer imprint is published by the registered company Springer Nature Switzerland AG
The registered company address is: Gewerbestrasse 11, 6330 Cham, Switzerland

Preface

This volume contains expanded versions of research papers presented at the international sessions of Annual Conference of the Japanese Society for Artificial Intelligence (JSAI), which was held online on June 2020. The JSAI annual conferences are considered key events for our organization, and the international sessions held at these conferences play a key role for the society in its efforts to share Japan's research on artificial intelligence with other countries. In recent years, AI research has proved of great interest to business people. The event draws both more and more presenters and attendees every year, including people of diverse backgrounds such as law and the social sciences, in additional to artificial intelligence. We are extremely pleased to publish this collection of papers as the research results of our international sessions.

The 2020 JSAI Annual Conference drew a record number of contributions, including 85 submissions for the international session, surpassing the number for the previous year. Contributors selected topics from the five categories of machine learning, human interface and education aid, knowledge engineering, agents, and robots and real worlds, with half of these pertaining to machine learning. The papers are selected for presentation at the international sessions over a period of two months, which at least two reviewers screen each paper. In the second phase of the review process, the chosen papers are once again evaluated for originality, significance, and quality. The final twenty-four are included in this book. All are original and high quality, and represent key contributions to AI research.

We would like to extend our deepest appreciation to President Itsuki Noda, Vice President Koji Morikawa (General Chair of JSAI2020), Ex-President Naohiko Uramoto, and Executive Committee Chair Akisato Kimura, as well as to the JSAI administrative staff and Springer publishing staff for their tremendous assistance on this project. We would also like to thank the authors of the papers contained in this book and all international session contributors.

January 2021

Katsutoshi Yada
Chief Editor

Daisuke Katagami
Program Chair, JSAI 2020

Takayuki Ito
Yasufumi Takama
Eri Sato-Shimokawara
Akinori Abe
Junichiro Mori
Naohiro Matsumura
Hisashi Kashima
International Program Committee, JSAI 2020

Organization

The 34th Annual Conference of the Japanese Society for Artificial Intelligence, 2020

Conference Chair

Koji Morikawa	Connect Japan

Executive Committee Chairs

Akisato Kimura	NTT, Japan
Takeo Hosomi (vice)	NEC, Japan
Takuma Otsuka (sub)	NTT, Japan

Program Committee Chair

Daisuke Katagami	Tokyo Polytechnic University, Japan
Mitsunori Matsushita (vice)	Kansai University, Japan
Takahito Tomoto (sub)	Tokyo Polytechnic University, Japan

Program Committee

Masahiro Ito	Toshiba, Japan
Hiroto Yonenoh	Nagoya University, Japan
Tomohiro Harada	Tokyo Metropolitan University, Japan
Satoshi Yoshida	NEC, Japan
Ken Tsutsuguchi	Sojo University, Japan

Syuuichi Ojima	Sojo University, Japan
Noriko Horibe	Sojo University, Japan
Motoki Amagasaki	Kumamoto University, Japan
Masato Kiyama	Kumamoto University, Japan
Gou Koutaki	Kumamoto University, Japan
Yuzuru Okajima	NEC, Japan
Koh Mitsuda	NTT, Japan
Kohei Miyaguchi	IBM Research, Tokyo, Japan
Ryo Iwaki	IBM Research, Tokyo, Japan
Junpei Komiyama	New York University, USA
Hidekazu Oiwa	Google, USA
Takashi Nishibayashi	Voyage Group, Japan
Takehiro Terada	Takushoku University, Japan
Kyota Tsutsumida	NTT, Japan
Yoichi Sasaki	NEC, Japan
Jun Ichikawa	Kanagawa University, Japan
Keisuke Fujii	Nagoya University, Japan
Junpei Tsuji	Toyo University, Japan
Tomoyuki Kajiwara	Osaka University, Japan
Dai Hasegawa	Hokkai Gakuen University, Japan
Kazuya Narita	Insight Tech, Japan
Hiromi Narimatsu	NTT, Japan
Naoko Hayashida	Fujitsu Laboratories, Japan
Masakazu Hirokawa	Tsukuba University, Japan
Yuki Ogawa	Ritsumeikan University, Japan
Megumi Kurayama	National Institute of Technology, Hakodate College, Japan
Jun Sugiura	Yahoo Japan, Japan
Masashi Okada	Panasonic, Japan
Shun Tanaka	Future Corporation, Japan
Satoshi Oyama	Hokkaido University, Japan
Taiki Todo	Kyushu University, Japan
Tomoki Aburatani	Osaka Prefecture University, Japan
Kento Koike	Tokyo Polytechnic University, Japan
Yasufumi Takama	Tokyo Metropolitan University, Japan
Eri Shimokawara	Tokyo Metropolitan University, Japan
Takayuki Ito	Nagoya Institute of Technology, Japan
Akinori Abe	Chiba University, Japan
Katsutoshi Yada	Kansai University, Japan
Junichiro Moi	Tokyo University, Japan
Naohiro Matsumura	Osaka University, Japan
Hisashi Kashima	Kyoto University, Japan
Lieu-Hen Chen	National Chi Nan University, Taiwan
Yihsin Ho	Takushoku University, Japan
Yuya Yoshikawa	Chiba Institute of Technology, Japan
Hirotoshi Taira	Osaka Institute of Technology, Japan

Contents

A Node Classification Approach for Dynamically Extracting the Structures of Online Discussions

Shota Suzuki[1(✉)], Takayuki Ito[2], Ahmed Moustafa[1], and Rafik Hadfi[1]

[1] Nagoya Institute of Technology, Gokiso-cho, Showa-ku, Nagoya-shi, Aichi-ken 466-8555, Japan
suzuki.shota@itolab.nitech.ac.jp

[2] Kyoto University, Yoshida Honmachi, Sakyo-ku, Kyoto-shi, Kyoto-fu 606-8501, Japan
http://www.itolab.nitech.ac.jp/itolab_hp_2017/

Abstract. Online discussion platforms require extracting the discussion structure in order to support understanding the flow of these discussions. Towards this end, this paper proposes an approach that performs node classification which is the first step towards extracting the structure of online discussions. In this regard, the proposed approach employs a graph attention network (GAT) in order to directly learn the discussion structure. In specific, the GAT, which is a type of graph neural networks (GNNs), encodes the graph structures directly. In addition, the GAT, which is based on attention architecture, is able to deal with different graph structures. In order to evaluate the proposed approach, we have conducted a set of experiments on the persuasive essay dataset that is styled using the issue-based information system (IBIS). The experimental results show that the proposed approach is able to classify the nodes in online discussion structures accurately.

Keywords: Agent · Deep learning · Facilitation · Natural language processing · Online discussion

1 Introduction

Online discussion support systems have received great attention because they are the next generation approaches for open and public citizen forums [10–12,17,22,28]. In this regard, intelligent consensus-building support systems, namely, COLLAGREE [10,11,22,28] and D-Agree [12] are developed and employed in several real-world experiments. In this context, COLLAGREE and D-Agree had provided support functions for human facilitators who coordinate, lead, integrate, classify, and summarize discussions in order to reach consensus. In their experiments, the consensus-building support systems successfully gathered many more opinions than comparable meetings and made the participants recognize the importance of facilitators.

K. Yada et al. (Eds.): JSAI 2020, AISC 1357, pp. 1–12, 2021.
https://doi.org/10.1007/978-3-030-73113-7_1

In the fields of online discussions and facilitation, our ultimate goal is to create an automated facilitation agent that takes the role of human facilitators. This agent has several benefits such as reacting quickly, facilitating the ongoing discussions in 24 h a day, and making fewer mistakes compared to human facilitators. Towards this end, we have developed the automated facilitation agent that can facilitate online discussions in COLLAGREE and D-Agree.

The automated facilitation agent needs a function that extracts the structures of online discussions in order to understand the situations of these discussions. Understanding the situations of online discussions enables the automated facilitation agent to post facilitations that coordinate and enhance the discussions. Although human facilitators can understand the situations by witnessing the discussions, the automated facilitation agent cannot witness the discussions. Hence, we address the problem of extracting the structures of online discussions.

The task of extracting online discussion structures consists of two subtasks which are node classification and link prediction [27]. The first subtask of node classification means classifying sentences which compose the submissions in online discussions. On the other hand, the second subtask of link prediction means predicting the relationships amongst classified nodes. In this paper, we challenge the first subtask of node classification because node classification with high accuracy results in link prediction with high accuracy.

Towards this end, we propose an approach that directly learns the graph structure for node classification as opposed to the existing feature-based approaches. The proposed approach employs a graph attention network (GAT) [29] that is a kind of graph neural networks (GNNs). In this regard, GAT and GNNs encode the graph structure directly using a neural network model. In particular, the GAT can deal with different graph structures because it employs attention mechanisms [1] unlike other GNNs. In this respect, the proposed approach can address many kinds of graph structures created from natural language texts.

In order to evaluate the proposed approach, we conduct a set of experiments on the persuasive essay dataset [26] formatted into the issue-based information system (IBIS) [15]. The IBIS is an approach for structuring online discussions. The elements of the IBIS are issues that are stated in the form of a controversial question, positions that are responses to an issue, and arguments that are evidence offered to support or oppose a position. The experimental results show achieving high F1 scores when classifying nodes into issues, positions, and arguments. As a result, we conclude that the proposed approach which directly learns the graph structure performs node classification well.

The remainder of this paper is organized as follows. In Sect. 2, we introduce the related works in the fields of the GNNs and the GAT. In Sect. 3, we detail the proposed approach which directly learns the graph structure with a GAT. In Sect. 4, we explain the experimental setup on the dataset of persuasive essays. In Sect. 5, we evaluate the experimental results. In Sect. 6, we discuss the classification performance of the proposed approach. In Sect. 7, we conclude this paper.

2 Related Works

2.1 Graph Neural Networks

GNNs encode graph structures directly using a neural network model. In this regard, GNNs had originated from [24] which showed that neural networks can deal with structured domains such as medical and technical diagnosis, molecular biology, chemistry, speech, text processing, and many others. In particular, the research [24] pointed out that standard neural networks and feature-based approaches are usually believed to be inadequate for dealing with complex structures represented as lists, trees, and graphs of variable sizes and complexities. Also, the research [24] mentioned that the current neural networks do not allow an efficient classification of structures of different sizes. In the attempts to adopt those complex structures, [8] presented the first GNN that is capable of directly processing graphs as opposed to feature-based approaches. In this respect, the research [8] introduced that GNNs can extend recursive neural networks by applying it on directed, undirected, labeled, and cyclic graphs. Particularly, the research indicated that valuable information is lost when traditional machine learning techniques try to cope with graphical data structures by a preprocessing procedure that transforms the graphs into simpler representations such as vectors or sequences of reals. Due to the lack of valuable information, these traditional machine learning techniques may suffer from poor performance and generalization. Besides, many approaches based on the GNNs such as a recurrent GNN [21], a convolutional GNN [3,9], a variational graph auto-encoder [14], and a spatio-temporal graph neural network [23] were proposed in order to promote performance and generalization in many fields of applications. In this regard, these approaches based on the GNNs have proven to learn many kinds of complex graph structures.

2.2 Graph Attention Network

The GAT [29] can deal with different graph structures because of its attention-based architecture. On the other hand, the usual GNNs cannot deal with different graph structures because they compute the eigendecomposition of the graph Laplacian, and their learned filters depend on the Laplacian eigenbasis in encoding graph structures [13]. As an advantage of dealing with different graph structures, the GAT achieved state-of-the-art performance across four established graph-based benchmark datasets of Cora, Citeseer, Pubmed, and protein-protein interaction (PPI). In particular, the results on the PPI dataset showed a significant improvement in the performance. In this regard, the research pointed out that the GAT has large predictive power by observing the entire neighborhood.

Although the existing research [29] adopted the tasks in the fields of citation network and biochemistry, the proposed approach tackles component classification in the field of argument mining. Therefore, it is rich in originality that the proposed approach performs a subtask in the field of argument mining. Besides,

Table 1. A text formatted into the IBIS

Issue1	Are cultural identities issues for immigrants?
Issue2	What is required of immigrants to the country which adopted them?
Position1	Keeping the cultural traditions in the destination countries is tremendous important
Argument1	They need a connection back to their country as well as teach their children their value of origin
Argument2	Fail to create this familiarity makes them felt isolated, in the extreme can lead to social disorder like autism
Issue3	What is required of immigrants to the country which adopted them?
Position2	Immigrants should follow the local customs in order to integrate into their adopted countries' cultures

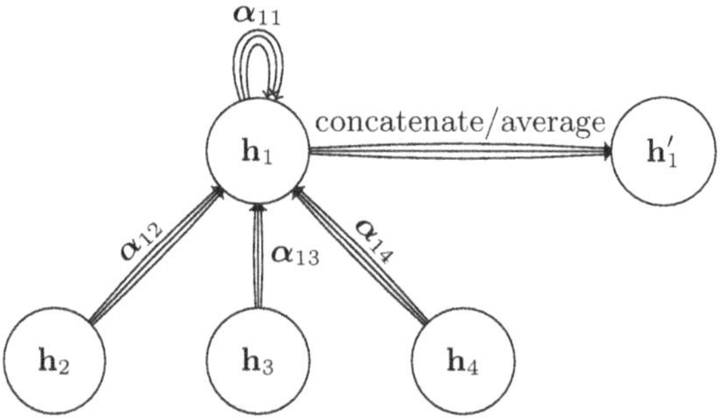

Fig. 1. The single layer of the proposed approach

though the existing research [29] applied the GAT for semi-supervised learning, the proposed approach applies the GAT for supervised learning for the component classification. Hence, it has great value that the results show that the proposed approach based on the GAT is also suitable for supervised learning.

3 The Proposed Approach

3.1 The Single GAT Layer

The GAT is a kind of graph neural networks (GNNs) which encode the graph structure directly using a neural network. In particular, the GAT can deal with different graph structures because of its attention mechanism. Following the original study [29], the layer compute a linear combination of the features in order to get the output of final features for every node:

$$\mathbf{h}_i' = \sigma\left(\sum_{j \in \mathcal{N}_i} \alpha_{ij} \mathbf{W} \mathbf{h}_j\right)$$

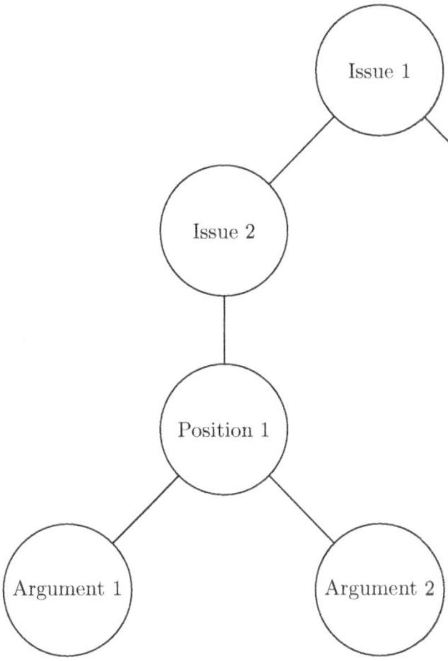

$$G = (V, E)$$

$$V = \{Issue\,1, Issue\,2, Issue\,3, Position\,1, Position\,2, Argument\,1, Argument\,2\}$$
$$E = \{(Issue\,1, Issue\,2), (Issue\,2, Position\,1), (Position\,1, Argument\,1), (Position\,1, Argument\,2), (Issue\,1, Issue\,3), (Issue\,3, Position\,2)$$

Fig. 2. The graph converted from the text

where σ is an activation function, $\mathcal{N}_i$ is first-order neighbors of node i including a self loop, α_{ij} is a element of an attention weight vector, $\mathbf{W}$ is a weight matrix for shared linear transformation, and $\mathbf{h}_j$ is features of node j. α_{ij} is calculated by the softmax function [7]:

$$\alpha_{ij} = \frac{\exp(a(\mathbf{W}\mathbf{h}_i, \mathbf{W}\mathbf{h}_j))}{\sum_{k \in \mathcal{N}_i} \exp(a(\mathbf{W}\mathbf{h}_i, \mathbf{W}\mathbf{h}_k))}$$

where $a : \mathbb{R}^{F'} \times \mathbb{R}^{F'} \rightarrow \mathbb{R}$ is an attention mechanism which calculates the importance of node j's features to node i's features. For the final layer, the proposed approach apply averaging of K independent attention mechanisms:

$$\mathbf{h}'_i = \sigma\left(\frac{1}{K}\sum_{k=1}^{K}\sum_{j \in \mathcal{N}_i} \alpha_{ij}^k \mathbf{W}^k \mathbf{h}_j\right).$$

As a result, the single layer of the proposed approach is illustrated as shown in Fig. 1.

3.2 Node Classification with Graph Attention Network

We propose an approach that directly learns the graph structure with the GAT. First, the proposed approach converts a text that consists of its sentences and their relations into a graph. Therefore, nodes of a graph correspond to sentences of a text, links of a graph correspond to relations of a text. For instance, there is a text formatted as the IBIS as shown in Table 1. In addition, the sentences in the text have relations such as {Issue 1 ↔ Issue 2, Issue 2 ↔ Position 1, Position 1 ↔ Argument 1, Position 1 ↔ Argument 2, Issue 1 ↔ Issue 3, Issue 3 ↔ Position 2}. And then, the proposed approach converts this text into the graph as shown in Fig. 2.

Second, the proposed approach embeds each node into a vector in order to build the features. In this regard, the proposed approach employs the below features as referred to in study [26]:

- Lexical features are binary unigram and 2K most frequent dependency word pairs.
- Indicator features are forward, backward, thesis, and rebuttal Indicators. In addition, the features contain binary values if first-person Indicators are present.
- Embedding features are word embeddings with the bidirectional encoder representations from transformers (BERT) [5].

At last, the proposed approach directly learns the structure of the graph G with supervised learning. In this regard, the proposed approach minimizes the cross-entropy loss in order to tune the weights of the hidden states of the model according to:

$$\mathcal{L} = -\sum_{i \in N} \sum_{c \in C} t_{ic} \ln h'_{ic}$$

where N is the number of sentences in an text, C is the set of classes of nodes in the IBIS ({Issue, Position, Argument}), e.g. $\{MajorClaim, Claim, Argument\}$, t_{ic} is ground truth of the i-th node and h'_{ic} is the output for the i-th node.

4 Experiments

4.1 Dataset

In order to evaluate the proposed approach, we conduct a set of experiments on the dataset of persuasive essays [26] formatted into the IBIS. In this regard, the dataset consists of 90 English texts. In order to evaluate the generalization performance, we randomly split the dataset into training data, validation data, and test data. The number of texts, the number of sentences, and the number of tokens are shown in Table 2. In addition, the classes of the sentences in this dataset are issues, positions, and arguments.

Table 2. The number of texts, number of sentences, and number of tokens in training/validation/test datasets.

	Training	Validation	Test
The number of texts	54	18	18
The number of sentences	1,238	420	413
The number of tokens	15,359	5,216	5,009
Average sentences/text	22.93	23.33	22.94
Average tokens/sentence	12.41	12.42	12.13

Table 3. The parameters for the proposed approach

The number of the GAT layers	2
The number of a layer dimensions	256
The dropout [25] rate of a GAT layer	0.1
The number of hidden attention heads	4
The number of output attention heads	4
The dropout [25] rate of an attention layer	0.1

4.2 Model Settings

In order to perform node classification with directly learning graph structures, we implement the proposed approach in Python. In this regard, we employ the PyTorch [18] which is an open-source machine learning framework that accelerates the path from research prototyping to production deployment. Besides, we adopted the deep graph library [30] which is a Python package built for easy implementation of graph neural network models family, on top of the existing deep learning frameworks such as PyTorch, MXNet, Gluon, and so on. In addition, we adopt the stochastic gradient descent [20] as an optimizer and the stochastic gradient descent with warm restarts (SGDR) [16] as a method for adjusting the learning rate. Moreover, the representative parameters of the model are shown in Table 3, and the representative parameters for training the model are shown in Table 4.

Table 4. The parameters for training the proposed approach

Epochs	2000
Batch size	18
Learning rate	0.001
Momentum for the SGD	0.9
The number of first restart for the SGDR	200

Table 5. The results of experiments

	Issues			Positions			Arguments		
	P	R	F	P	R	F	P	R	F
SVM	1.000	0.630	0.773	1.000	0.196	0.327	0.680	1.000	0.809
RF	0.967	0.967	0.967	0.897	0.380	0.534	0.798	0.983	0.881
GB	0.968	1.000	0.984	0.737	0.609	0.667	0.860	0.908	0.883
MLP	0.979	1.000	**0.989**	0.654	0.761	0.704	0.906	0.838	0.871
The proposed approach	0.936	0.957	0.946	0.910	0.880	**0.895**	0.978	0.983	**0.980**

4.3 Baselines

The proposed approach is compared with several baselines of the existing feature-based approaches. Particularly, we perform the same task with four machine learning approaches such as support vector machine (SVM) [4], random forest (RF) [2], gradient boosting (GB) [6], and multilayer perceptron (MLP).

4.4 Evaluation Methods

We evaluate the proposed approach and the baselines with the values of precision, recall, and F1 score [19]. In this regard, F1 score is the harmonic average of precision and recall:

$$F1\ Score = \frac{2}{\frac{1}{Precision} + \frac{1}{Recall}}.$$

5 Results

The results of the experiments are shown in Table 5. The results show that the proposed approach achieves F1 score values of 0.946 in issues classification, 0.895 in positions classification, 0.980 in arguments classification.

To summarize, the results show that the proposed approach which directly learns the graph structure performs node classification well. Particularly, the results of classifying nodes into positions and arguments show the higher F1 score value than the baselines. In this regard, this result is attributed to the fact that the proposed approach learns the graph structures. Particularly, the proposed approach is useful for classifying nodes into positions because the nodes of positions have relations with both of the nodes of issues and arguments. On the other hand, the results of classifying nodes into issues show a little lower F1 score value than the baselines of the RF, GB, MLP. Therefore, we should employ features that are efficient for the proposed approach in order to classify nodes into issues more accurately.

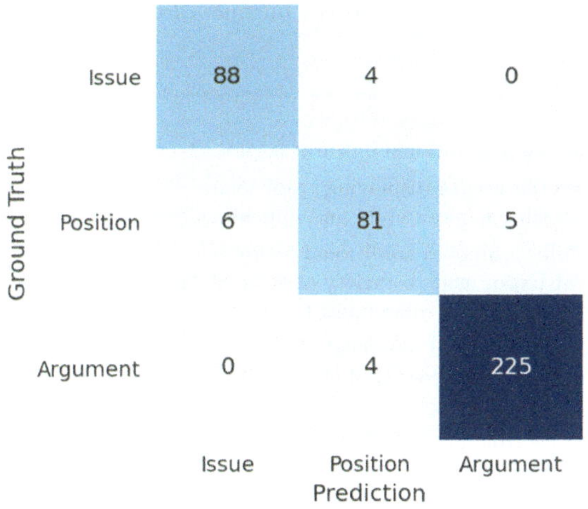

Fig. 3. Confusion matrix.

Table 6. The result of feature ablation.

Features	Issues			Positions			Arguments		
	P	R	F	P	R	F	P	R	F
All	0.936	0.957	0.946	0.910	0.880	0.895	0.978	0.983	0.980
All - Indicator	0.945	0.935	0.940	0.880	0.880	0.880	0.974	0.978	0.976
All - Lexical	0.926	0.946	0.935	0.886	0.848	0.867	0.970	0.978	0.974
All - Embedding	0.659	0.630	0.644	0.500	0.163	0.246	0.742	0.956	0.836

6 Discussion

6.1 Performance for Classifying Each Type of Node

In order to analyze the classification performance of the proposed approach, Fig. 3 shows the confusion matrix of the experimental results in Sect. 5. Figure 3 indicates that the proposed approach classifies issue and argument nodes almost exactly. In contrast, there is room for improvement in classifying claim nodes. In order to improve the performance more, it is required for the proposed approach to employ additional efficient features for node classification.

6.2 Feature Ablation Study

We examine which feature is efficient for node classification with a set of feature ablation experiments. In this respect, we except one of three kinds of features and conduct experiments with the rest features. The results of the ablation experiments are shown in Table 6. Table 6 demonstrates that the embedding feature is the most efficient one of three kinds of features. On the other hand, the indicator feature and the lexical feature improve the classification performance of the proposed approach a little.

Table 7. An example text and the outputs of the proposed approach.

Text	Prediction	Ground truth
What is the role of technology in tradition?	Issue	Issue
How can technology help traditional cultural heritage?	Issue	Issue
Technology enriches the way of displaying traditional cultural heritage, making it more vivid and appealing	Position	Position
China has successfully promoted traditional techniques in the Shanghai World Expo, using a variety of state-of-the-art technological methods, such as robots and LED screens	Argument	Argument
Many tourists around the world marveled at the perfect synthesis of the modern technology and the traditional culture	Argument	Argument
Technology plays a key role in promoting traditional techniques and lifestyles	Position	Position
What is the possible problem that technology bring for tradition?	Issue	Issue
Technology may have a detrimental effect on the traditional lifestyles	Position	Position
People, especially youngsters, are crazier about fresh and advanced things, such as digital products, thus becoming indifferent to traditional techniques	Argument	Argument
People tend to follow the trend back to the tradition in the recent years	Argument	Argument
What can technology do for tradition techniques and lifestyles?	Issue	Issue
Technology supports the preservation and promotion of traditional techniques and lifestyles	Position	Position
Only when traditional culture is integrated with the modern technology can it be developed in the long run	Argument	Argument
What do you think about the opinion that technology will destroy the tradition?	Issue	Issue
Their concern is groundless	Issue	Position

6.3 Case Study

In order to validate the proposed approach, we observe the outputs of the proposed approach. Table 7 shows an example text and the outputs of the proposed approach. We confirmed that the proposed approach classified all issue nodes and all argument nodes accurately. On the other hand, the proposed approach classified a position node of "Their concern is groundless." into an issue node. The additional efficient features would help the proposed approach classify the node accurately.

7 Conclusion and Future Work

This paper proposes a novel approach that classifies discussion nodes via directly learning the graph structures. In this regard, the proposed approach employs a special type of graph neural network, namely, graph attention network. In order to evaluate the proposed approach, a set of experiments has been conducted on the persuasive essay dataset that was formatted into IBIS structure. The experimental results show higher F1 scores when classifying nodes into positions and arguments as compared to other approaches. Future work is set to improve the accuracy of the issue nodes classification. In addition, devising efficient approaches for link prediction is also set for future work.

Acknowledgements. This work was supported by JST CREST Grant Number JPMJCR15E1 including AIP challenge program, Japan.

References

1. Bahdanau, D., Cho, K., Bengio, Y.: Neural machine translation by jointly learning to align and translate. arXiv preprint arXiv:1409.0473 (2014)
2. Breiman, L.: Random forests. Mach. Learn. **45**(1), 5–32 (2001)
3. Bruna, J., Zaremba, W., Szlam, A., Lecun, Y.: Spectral networks and locally connected networks on graphs. In: International Conference on Learning Representations (ICLR 2014), CBLS, April 2014 (2014)
4. Cortes, C., Vapnik, V.: Support-vector networks. Mach. Learn. **20**(3), 273–297 (1995)
5. Devlin, J., Chang, M.W., Lee, K., Toutanova, K.: Bert: pre-training of deep bidirectional transformers for language understanding. arXiv preprint arXiv:1810.04805 (2018)
6. Friedman, J.H.: Greedy function approximation: a gradient boosting machine. Ann. Stat. 1189–1232 (2001)
7. Goodfellow, I., Bengio, Y., Courville, A.: Deep Learning. MIT Press, Cambridge (2016)
8. Gori, M., Monfardini, G., Scarselli, F.: A new model for learning in graph domains. In: Proceedings of the 2005 IEEE International Joint Conference on Neural Networks, vol. 2, pp. 729–734. IEEE (2005)
9. Henaff, M., Bruna, J., LeCun, Y.: Deep convolutional networks on graph-structured data. arXiv preprint arXiv:1506.05163 (2015)
10. Ito, T., Imi, Y., Ito, T., Hideshima, E.: COLLAGREE: a faciliator-mediated large-scale consensus support system. Collective Intell. **2014**, 10–12 (2014)
11. Ito, T., Imi, Y., Sato, M., Ito, T., Hideshima, E.: Incentive mechanism for managing large-scale internet-based discussions on COLLAGREE. Collective Intell. **2015** (2015)
12. Ito, T., Suzuki, S., Yamaguchi, N., Nishida, T., Hiraishi, K., Yoshino, K.: D-agree: crowd discussion support system based on automated facilitation agent. In: AAAI, pp. 13614–13615 (2020)
13. Kipf, T.N., Welling, M.: Semi-supervised classification with graph convolutional networks. arXiv preprint arXiv:1609.02907 (2016)
14. Kipf, T.N., Welling, M.: Variational graph auto-encoders. Stat **1050**, 21 (2016)

15. Kunz, W.: Issues as elements of information systems. Institute of Urban & Regional Development, University of California Working Paper, Berkeley 131 (1970)
16. Loshchilov, I., Hutter, F.: SGDR: stochastic gradient descent with warm restarts. arXiv preprint arXiv:1608.03983 (2016)
17. Malone, T.W.: Superminds: the surprising power of people and computers thinking together. Little, Brown Spark (2018)
18. Paszke, A., et al.: Pytorch: an imperative style, high-performance deep learning library. In: Wallach, H., Larochelle, H., Beygelzimer, A., d' Alché-Buc, F., Fox, E., Garnett, R. (eds.) Advances in Neural Information Processing Systems 32, pp. 8024–8035. Curran Associates, Inc. (2019). http://papers.neurips.cc/paper/9015-pytorch-an-imperative-style-high-performance-deep-learning-library.pdf
19. Powers, D.M.: Evaluation: from precision, recall and f-measure to ROC, informedness, markedness and correlation (2011)
20. Robbins, H., Monro, S.: A stochastic approximation method. Ann. Math. Stat. 400–407 (1951)
21. Scarselli, F., Gori, M., Tsoi, A.C., Hagenbuchner, M., Monfardini, G.: The graph neural network model. IEEE Trans. Neural Netw. **20**(1), 61–80 (2008)
22. Sengoku, A., et al.: Discussion tree for managing large-scale internet-based discussions. Collective Intell. **2016** (2016)
23. Seo, Y., Defferrard, M., Vandergheynst, P., Bresson, X.: Structured sequence modeling with graph convolutional recurrent networks. In: Cheng, L., Leung, A.C.S., Ozawa, S. (eds.) ICONIP 2018. LNCS, vol. 11301, pp. 362–373. Springer, Cham (2018). https://doi.org/10.1007/978-3-030-04167-0_33
24. Sperduti, A., Starita, A.: Supervised neural networks for the classification of structures. IEEE Trans. Neural Netw. **8**(3), 714–735 (1997)
25. Srivastava, N., Hinton, G., Krizhevsky, A., Sutskever, I., Salakhutdinov, R.: Dropout: a simple way to prevent neural networks from overfitting. J. Mach. Learn. Res. **15**(1), 1929–1958 (2014)
26. Stab, C., Gurevych, I.: Parsing argumentation structures in persuasive essays. Comput. Linguist. **43**(3), 619–659 (2017)
27. Suzuki, S., et al.: Extraction of online discussion structures for automated facilitation agent. In: Ohsawa, Y., et al. (eds.) JSAI 2019. AISC, vol. 1128, pp. 150–161. Springer, Cham (2020). https://doi.org/10.1007/978-3-030-39878-1_14
28. Takahashi, K., et al.: Incentive mechanism based on qualit of opinion for large-scale discussion support. Collective Intell. **2016**, 16 (2016)
29. Veličković, P., Cucurull, G., Casanova, A., Romero, A., Lio, P., Bengio, Y.: Graph attention networks. arXiv preprint arXiv:1710.10903 (2017)
30. Wang, M., et al.: Deep graph library: towards efficient and scalable deep learning on graphs. arXiv preprint arXiv:1909.01315 (2019)

Visualizing Road Condition Information by Applying the AutoEncoder to Wheelchair Sensing Data for Road Barrier Assessment

Goh Sato[1(✉)], Takumi Watanabe[1], Hiroki Takahashi[1], Yojiro Yano[1], Yusuke Iwasawa[2], and Ikuko Eguchi Yairi[1]

[1] Graduate School of Science and Engineering, Sophia University, 7-1 Kioicho, Chiyoda-ku, Tokyo 102-8554, Japan
sato.g@yairilab.net

[2] Graduate School of Technology Management for Innovation, The University of Tokyo, 7-3-1 Hongo, Bunkyo-ku, Tokyo 113-8656, Japan

Abstract. Providing accessibility information about sidewalks for people with difficulties with moving is an important social issue. Visualizing road surface conditions to show the accessibilities of the road is effective for this issue. However, conventional methods of collecting huge area accessibility information are based on manpower and have the problem of the large costs of time and money. To solve this problem, we have been proposing and implementing a system for estimating road surface conditions by machine learning with measured values of an acceleration sensor attached to wheelchairs. This paper examined the appropriateness of reconstruction errors which are calculated by Convolutional Variational AutoEncoder to assess the degree of road burden suitable for each user. The evaluation was conducted by calculating reconstruction errors from the traveling data of 14 wheelchair users and creating a map that reflects the information of the calculated errors. The evaluation results suggest that reconstruction errors can reflect the degree of the burden on each wheelchair user during traveling.

Keywords: Variational autoencoder · Convolutional neural network · Sidewalk accessibility · Human activity recognition

1 Introduction

Ensuring the accessibility of the sidewalk is an important social issue for supporting the mobility of the elderly and the mobility impaired, such as wheelchair users. One solution to this issue using information and communication technologies is to develop an accessibility map as an extensive geographic information system (GIS) that provides accessibility information on sidewalks [1, 2]. The personalized accessibility map (PAM) was also proposed as an optimal route navigation system for each user with difficulties in moving [3]. The basic idea of these systems is collecting and digitizing road conditions as accessibility information and visually providing the collected information to users. This paper, hereafter, calls whole process of this idea 'visualization'.

K. Yada et al. (Eds.): JSAI 2020, AISC 1357, pp. 13–24, 2021.
https://doi.org/10.1007/978-3-030-73113-7_2

The main problem to be solved for the visualization is a large-scale data collection of sidewalk accessibility information. Conventional methods of the data collection are as follows: experts evaluate each case from images taken of sidewalks [4], and cooperator evaluate sidewalks by crowdsourcing [5, 6]. These methods are based on manpower and have a problem that collecting large-scale information is difficult because of the time and money costs. Recently, due to the further development of intelligent gadgets, such as smartphones and wristwatch-shaped vital sensors, the ability to sense human activities is improving [7, 8]. To solve the problems above, we have proposed and implemented a system for estimating and classifying road surface conditions by using machine learning from the measured values of acceleration sensors attached to wheelchairs [9–13]. These methods have the difficulty of preparing enough labeled acceleration data and collecting many acceleration data for various road surface irregularities which are difficult for wheelchair users to travel, such as curbs and tactile blocks on the road. This paper uses Convolutional Variational AutoEncoder (CVAE), which has the advantage to detect anomalies even when labeled training data is none or low and anomaly data is low and imbalance. We examine whether the reconstruction error calculated by CVAE can be used as the degree of road burden on wheelchair users and try to provide road information suitable for each user.

2 Related Work

High-resolution satellite images were used for land cover classification according to the physical condition of the ground surfaces such as agricultural land, grazing land, and barren areas [14–16]. Eriksson et al. applied a simple machine learning approach to detect potholes on road surfaces using accelerometer and geographical positioning system (GPS) sensor data of taxies [17]. Allouch et al. applied machine learning to detect depressions in road surfaces using the accelerometer and gyroscope data of vehicles [18]. Abnormal traffic conditions in cities were detected using rich sensor data such as accelerometer, GPS, and voice data from smartphones [19, 20]. These methods can recognize large areas such as agricultural lands or roadways, but it is difficult to recognize conditions on sidewalks.

AutoEncoder (AE) is used for anomaly detection such as detection of anomaly events in video data [21], detection of breast cancer nuclei in medical images [22], and classification of human activities by using data with smartphones and wearable sensors [23, 24]. Generative Adversarial Network (GAN) is used for anomaly detection such as time-series sensor data [25] and real-world cyber-physical systems [26]. Metric Learning is used for anomaly detection from hyperspectral images [27]. Various models are used for anomaly detection, and Variational AutoEncoder (VAE) is used as well. VAE is good at modeling data uncertainty and detecting anomalies by mapping the characteristics of training data to the distribution of latent variables. VAE is used for anomaly detection such as daily activities [28], electrocardiogram [29], and multimodal sensory signals in combination with LSTM [30]. To the best of our knowledge, no research examines that the reconstruction error calculated by CVAE can be used as the degree of road burden for wheelchair users.

3 Overview of the Proposed Method

3.1 Acquisition of Wheelchair Sensing Data

Wheelchair sensing data used in this study is acceleration data of wheelchair users traveling about 1.4 km around Yotsuya Station in Tokyo (Route 1). The wheelchair users who participated in the experiment were six manual wheelchair users and three electric wheelchair users. The acceleration sensor (iPod touch) was attached under the wheelchair seat. Wheelchair users traveled Route 1 three laps. On the second lap, they traveled in the opposite direction from the first and third laps. Acceleration data were sampled in three directions of x, y, and z-axis at 50 Hz, and acceleration samples of about 7.5 h were obtained. To confirm the situations of the location where the acceleration samples were obtained, videos of the condition of wheelchair users and the road were taken during the experiment.

3.2 AutoEncoder and the Reconstruction Error

The AE compresses input data into small feature vectors and outputs reconstructed data which are similar to the input data. AE learns to minimize the errors between the reconstructed data and the input data. Data that resembles a lot of learned data can be accurately reconstructed. Conversely, data that does not resemble a lot of learned data are not accurately reconstructed. CVAE is a network in which the input data is convoluted into a latent variable z by the encoder; z is made into a standard normal distribution and deconvoluted by the decoder.

In this paper, the kernel size (number of feature maps) of the encoder is 5×1 (50), 5×1 (40), 3×1 (20), and the maximum pooling is 1×2, 1×2, 1×1. The kernel size of the decoder is 3×1 (20), 5×1 (40), 5×1 (50), 3×3 (1), and the upsampling is 1×1, 1×2, 1×2. The activation function is ReLU function. In the output layer, the reconstructed acceleration data is output linearly by using a Linear function. Adam is used for training the network, and the mean squared error is used as the loss function.

The CVAE model is created for each user. The acceleration data of eight of nine users are used for the model training, and the reconstruction error is calculated from the acceleration data of the remaining one. As a result, reconstruction errors corresponding to each of the nine wheelchair users are obtained. The reconstruction error is the mean squared error between the input and output data.

3.3 Visualizing Road Information

By using the reconstruction error and the corresponding location information, the information of the reconstruction error is reflected on Google Map (hereinafter referred to as the reconstruction error map). The color of plot points is determined by the ratio of each reconstruction error value to the highest reconstruction error value in three laps. Following the color bar shown in Fig. 1, the higher the ratio, the closer to red, and the lower the ratio, the closer to blue.

Reconstruction error maps we created are as follows: a reconstruction error map for each user (hereinafter referred to as map for each user) and the maximum reconstruction

Fig. 1. Color bar used to determine the color of points on the reconstruction error map

error map (hereinafter referred to as maximum map). The maximum map summarizes information of maps for all users. The method of creating the maximum map is as follows. First, the map of Route 1 is divided into grids. Second, in each grid, we search the maximum reconstruction error among reconstruction errors belonging to the grid. Finally, the point is plotted at each grid.

4 Evaluation of Visualizing Specific Road Information

4.1 Evaluation of a Reconstruction Error Map for Each User

This section evaluates a reconstruction error map for each user. Figure 2 shows the locations of curbs and tactile blocks (hereinafter referred to as anomaly road surfaces) on Route 1 where wheelchair users crossed during the experiment. In a map for each user, plot patterns are divided into two main types. Figure 3 shows an example of each pattern.

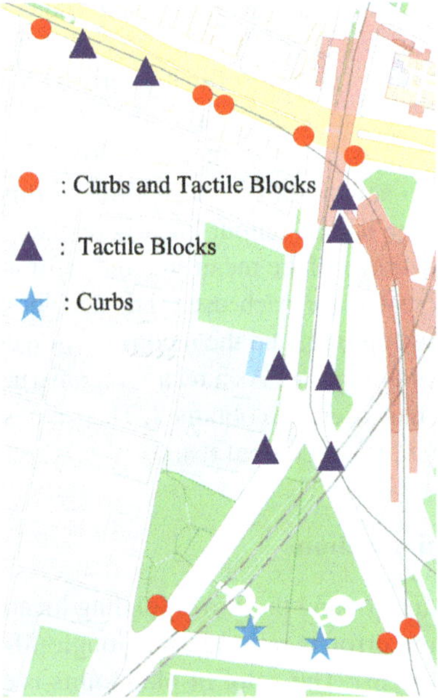

Fig. 2. Anomaly road surfaces existing on Route 1

The plotted pattern of the left figure is pattern 1 and the right figure is pattern 2. In pattern 1, reconstruction errors are high mainly on anomaly road surfaces. In pattern 2,

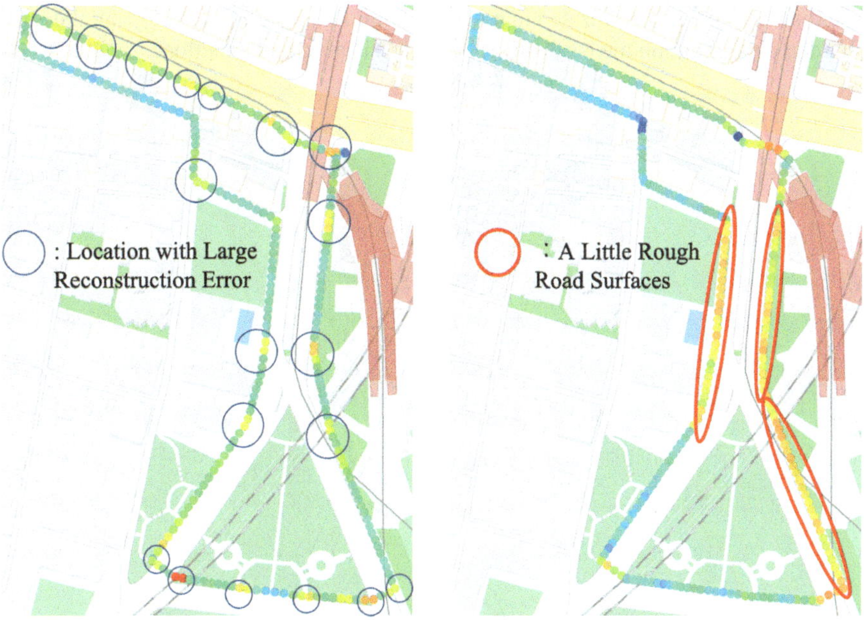

Fig. 3. A map for each user with high reconstruction error on the anomaly road surfaces (left). A map for each user with high reconstruction error on a little rough road surfaces (right)

the reconstruction errors are high mainly on a little rough road surfaces. The differences between wheelchair users belonging to each pattern are as follows: the wheelchair tires of users belonging to pattern 1 were thick, and the wheelchair tires of users belonging to pattern 2 were thin. Thin tires are more likely to vibrate than thick tires, even on a little rough road surfaces, which can be a burden on the wheelchair users who use a thin tire. These results indicate that a map for each user can reflect differences in the degree of burden depending on the wheelchair for each user.

4.2 Evaluation of the Maximum Reconstruction Error Map

Figure 4 shows the maximum map for Route 1. In the maximum map in Fig. 4, numbers are assigned to specific routes for explanations. As can be seen from the Fig. 2 and 4, reconstruction errors are relatively high on many anomaly road surfaces. Reconstruction errors are also high at locations other than anomaly road surfaces, such as corners.

The routes with mainly low reconstruction errors are 4, 5, 9, and 11. In these routes, the road surface conditions are smooth, and little traffic exists. Even if a little traffic exists, they are hard to be a burden on wheelchair users to travel because the road width is large. The routes with mainly moderate reconstruction errors are 2, 3, 6 and 7. In these routes, the road surface conditions are a little rough or sloped, and a little traffic exists. The routes with mainly high reconstruction errors are 1, 8 and 10. In these routes, the road surface conditions are smooth. However, traveling these routes are quite burdened on wheelchair users due to road situation, such as heavy traffic and narrow road width.

From these results, it is confirmed that the maximum map would reflect general degrees of burden depending on the road surface conditions and road situations.

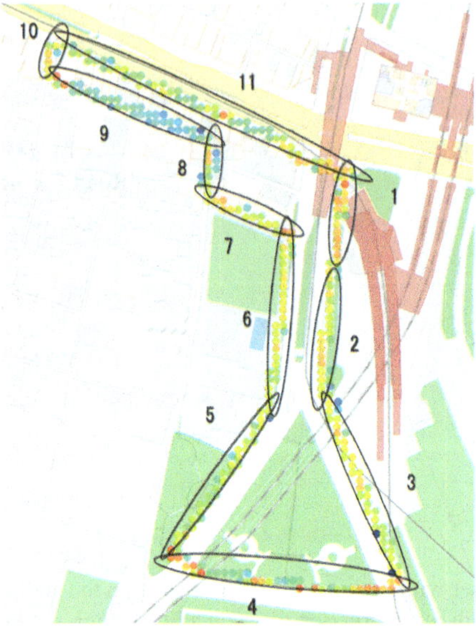

Fig. 4. The maximum map with numbers assigned to each route for explanations

5 Applicability of the Proposed Method to Other Routes

In the same way as the experimental method described in Sect. 3.1, the new experiment was conducted in December 2019 with a total of five participants, one manual and four electric wheelchair users, on a different route from Route 1. The wheelchair users traveled two rounds around the route which is about 800 m one way around Yotsuya Station (hereinafter referred to as Route 2). The acceleration data for Route 2 is input to the CVAE model that learned all acceleration data for Route 1, and maps for each user and a maximum map on Route 2 are created. We examine whether the CVAE model which is trained with the data of Route 1 has versatility by these maps. Figure 5 shows anomaly road surfaces on Route 2 where wheelchair users crossed during the experiment.

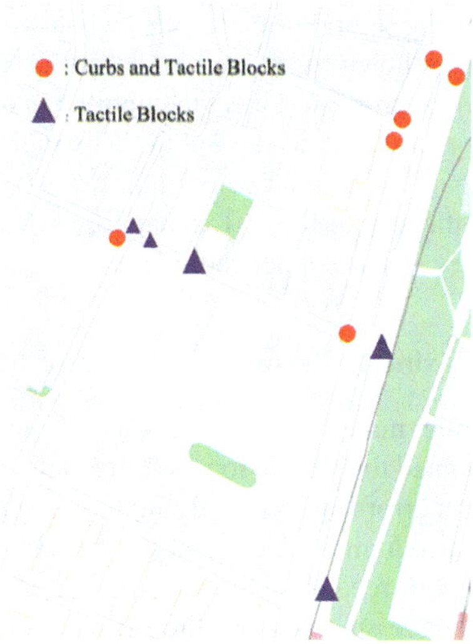

Fig. 5. Anomaly road surfaces existing on Route 2

5.1 Evaluation of a Reconstruction Error Map for Each User

In a map for each user on Route 2, the reconstruction errors are high on many anomaly road surfaces shown in Fig. 5. Plot patterns of maps for each user on Route 2 are also divided into two main types. Figure 6 shows an example of each pattern. The left figure is a map for each user of an electric wheelchair user, and the right figure is one of the manual wheelchair users.

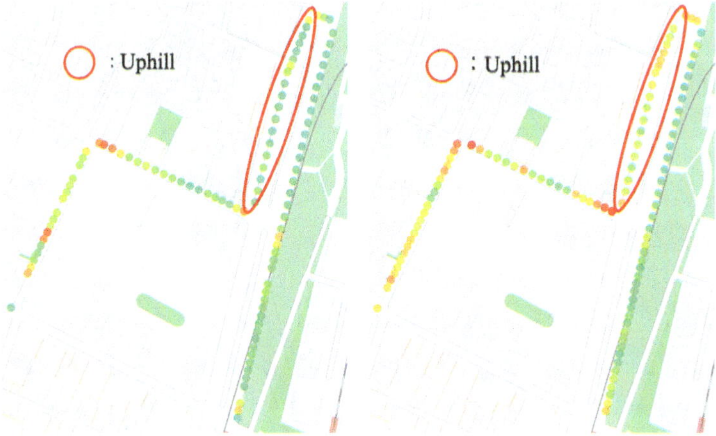

Fig. 6. A map for each user on Route 2 of an electric wheelchair user (left). A map for each user on Route 2 of a manual wheelchair user (right)

Both the left figure and the right figure are maps when wheelchair users traveled on the outward route. The area shown by the red circle in Fig. 6 is one of the uphills. In the left figure, reconstruction errors are low at the uphill. On the other hand, in the right figure, reconstruction errors are high at the uphill. This indicates a map for each user can detect the uphill that burdens manual wheelchair users but does not burden electric wheelchair users. As with the map of Route 1, a map for each user on Route 2 can reflect the differences in the degree of burden depending on the wheelchair of each user.

5.2 Evaluation of the Maximum Reconstruction Error Map

Figure 7 shows the maximum map for Route 2. As with Route 1, in the maximum map in Fig. 7, numbers are assigned to specific routes for explanations. As can be seen from Fig. 5 and 7, reconstruction errors are relatively high on many anomaly road surfaces. Reconstruction errors are also high at places that are raised by tree roots.

The routes with mainly low reconstruction errors are 1, 2, 3, 5 and 7. In these routes, the road surface conditions are smooth, and little traffic exists. Even if a little traffic exists, they are hard to be a burden on wheelchair users traveling because the road width is wide. The routes with mainly moderate reconstruction errors are 8, 9, and 11. In these routes, the road surface conditions are smooth, and a little traffic exists. A little traffic likely burdens wheelchair users due to the narrow road width. The routes with mainly high reconstruction errors are 4, 6 and 10. These routes are pedestrian crossing or a sharp curve with rough road surfaces. These results show that, as with the map of Route 1, the maximum map on Route 2 is able to visualize general degrees of burden depending on the road surface conditions and road situations.

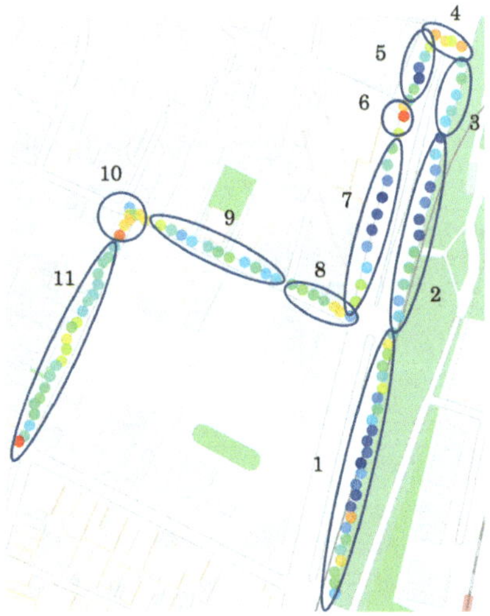

Fig. 7. The maximum map on Route 2 with numbers assigned to each route for explanations

6 Discussion

Generally, AE learns only normal data to detect anomalies. This paper used not only normal data but also anomaly data. Even so, reconstruction errors are high at anomaly road conditions. This is because, in the training data, the amount of data on anomaly road conditions is relatively small compared to the amount of data on normal road conditions. Some maps for each user could not be created successfully. For example, In Fig. 8, reconstruction errors are high on all routes. The user of this figure uses the Whill-model-A (Whill), which is a wheelchair with a different driving system design from a normal wheelchair. The map could not be created well because the driving characteristics are significantly different due to the performance difference between ordinary wheelchairs and Whill. We showed examples of maps for each user in Fig. 3 and 6. However, the output pattern of a map for each user varies depending on the person and the situation. Among the various output patterns, this paper presented the clear differences between users and showed that a map for each user can reflect each user's differences. To examine if a map for each user can reflect more detailed differences between each user, we need to ask for opinions to wheelchair users because the burden is subjective for each user.

Fig. 8. Example of a map for each user that could not be created successfully

7 Conclusion

In this paper, we examined if a map for each user could provide accessibility information suitable for each wheelchair user, and the maximum map could provide general accessibility information for all wheelchair users. We demonstrated a large amount of accessibility information can be obtained automatically by using the reconstruction error calculated by CVAE.

The CVAE model could not successfully reconstruct the acceleration data of the wheelchair user who didn't have traveling characteristics similar to users whose data was used for the model training. In order to create the CVAE model that calculates reconstruction errors suitable for more users, we will increase the number of acceleration data and improve the level of diversity of acceleration data. If the CVAE model is created by training data only for a certain wheelchair user, the reconstruction error calculated by that model should provide the more appropriate degree of the burden on that user. In order to examine this model, we will increase the number of acceleration data of a specific wheelchair user. We subjectively determined where is the burden on wheelchair users. These places may not burden on wheelchair users. We will examine the factors that are the burden on wheelchair users by conducting questionnaires to wheelchair users.

We used CVAE for anomaly detection in this paper. CVAE can reduce the input data to a low-dimensional latent variable z. Semi-supervised learning can improve accuracy by using the latent variable z [31]. In the future, we will also consider the classification of road conditions by using a semi-supervised learning combination with VAE.

Acknowledgments. We would like to thank all participants who helped to collect the sensing data. This study was supported by Tateishi Science and Technology Foundation in FY 2011–2012, the research grant by Chiyoda-ku (CHIYODAGAKU) in FY 2014–2016, and JSPS KAKENHI Grant-in-Aid for Scientific Research (B) Number 17H01946 and 20H04476.

References

1. Laakso, M., Sarjakoski, T., Sarjakoski, L.T.: Improving accessibility information in pedestrian maps and databases. Cartographica Int. J. Geogr. Inf. Geovisualization **46**(2), 101–108 (2011). https://doi.org/10.3138/carto.46.2.101
2. Matthews, H., Beale, L., Picton, P., Briggs, D.: Modelling access with GIS in urban systems (MAGUS): capturing the experiences of wheelchair users. Area **35**(1), 34–45 (2003). https://doi.org/10.1111/1475-4762.00108
3. Karimi, H.A., Zhang, L., Benner, J.G.: Personalized accessibility map (PAM): a novel assisted wayfinding approach for people with disabilities. Ann. GIS **20**(2), 99–108 (2014). https://doi.org/10.1080/19475683.2014.904438
4. Ponsard, C., Snoeck, V.: Objective accessibility assessment of public infrastructures. In: Miesenberger, K., Klaus, J., Zagler, W.L., Karshmer, A.I. (eds.) ICCHP 2006. LNCS, vol. 4061, pp. 314–321. Springer, Heidelberg (2006). https://doi.org/10.1007/11788713_47
5. Hara, K.: Scalable methods to collect and visualize sidewalk accessibility data for people with mobility impairments. In: Proceedings of the Adjunct Publication of the 27th Annual ACM Symposium on User Interface Software and Technology, Honolulu, HI, USA, pp. 1–4 (2014). https://doi.org/10.1145/2658779.2661163

6. Cardonha, C., Gallo, D., Avegliano, P., Herrmann, R., Koch, F., Borger, S.: A crowdsourcing platform for the construction of accessibility maps. In: Proceedings of the 10th International Cross-Disciplinary Conference on Web Accessibility, Rio de Janeiro, Brazil, p. 26 (2013). https://doi.org/10.1145/2461121.2461129
7. Swan, M.: The quantified self: fundamental disruption in big data science and biological discovery. Big Data **1**(2), 85–99 (2013). https://doi.org/10.1089/big.2012.0002
8. Nagamine, K., Iwasawa, Y., Matsuo, Y., Yairi, I.E.: An estimation of wheelchair user's muscle fatigue by accelerometers on smart devices. In: Adjunct Proceedings of the 2015 ACM International Joint Conference on Pervasive and Ubiquitous Computing and Proceedings of the 2015 ACM International Symposium on Wearable Computers, UbiComp/ISWC 2015 Adjunct, pp. 57–60. ACM, New York (2015). https://doi.org/10.1145/2800835.2800864
9. Iwasawa, Y., Yairi, I.E.: Life-logging of wheelchair driving on web maps for visualizing potential accidents and incidents. In: Anthony, P., Ishizuka, M., Lukose, D. (eds.) PRICAI 2012. LNCS (LNAI), vol. 7458, pp. 157–169. Springer, Heidelberg (2012). https://doi.org/10.1007/978-3-642-32695-0_16
10. Iwasawa, Y., Nagamine, K., Matsuo, Y., Eguchi Yairi, I.: Road sensing: personal sensing and machine learning for development of large scale accessibility map. In: Proceedings of the 17th International ACM SIGACCESS Conference on Computers & Accessibility, pp. 335–336. ACM (2015). https://doi.org/10.1145/2700648.2811366
11. Iwasawa, Y., Yairi, I.E., Matsuo, Y.: Combining, human action sensing of wheelchair users and machine learning for autonomous accessibility data collection. IEICE Trans. Inf. Syst. **E99-D**(4), 115–124 (2016). https://doi.org/10.1587/transinf.2015EDP7278
12. Yairi, I., et al.: Estimating spatiotemporal information from behavioral sensing data of wheelchair users by machine learning technologies. Information **10**(3), 114 (2019). https://doi.org/10.3390/info10030114
13. Watanabe, T., et al.: Weakly supervised learning for evaluating road surface condition from wheelchair driving data. Information **11**(1), 2 (2019). https://doi.org/10.3390/info11010002
14. Demir, I., et al.: Deepglobe 2018: a challenge to parse the earth through satellite images. In: Proceedings of the IEEE Conference on Computer Vision and Pattern Recognition (CVPR) Workshops, Salt Lake City, UT, USA, 18–22 June, pp. 172–181 (2018). https://doi.org/10.1109/CVPRW.2018.00031
15. Kuo, T.S., Tseng, K.S., Yan, J.W., Liu, Y.C., Frank Wang, Y.C.: Deep aggregation net for land cover classification. In: Proceedings of the IEEE Conference on Computer Vision and Pattern Recognition (CVPR) Workshops, Salt Lake City, UT, USA, 18–22 June, pp. 252–256 (2018). https://doi.org/10.1109/CVPRW.2018.00046
16. Rakhlin, A., Davydow, A., Nikolenko, S.: Land cover classification from satellite imagery with U-Net and Lovász-Softmax loss. In: Proceedings of the IEEE Conference on Computer Vision and Pattern Recognition (CVPR) Workshops, Salt Lake City, UT, USA, 18–22 June, pp. 262–266 (2018). https://doi.org/10.1109/CVPRW.2018.00048
17. Eriksson, J., Girod, L., Hull, B., Newton, R., Madden, S., Balakrishnan, H.: The pothole patrol: Using a mobile sensor network for road surface monitoring. In: Proceedings of the 6th International Conference on Mobile Systems, Applications, and Services, Breckenridge, CO, USA, 17–20 June 2008, pp. 29–39. ACM, New York (2008). https://doi.org/10.1145/1378600.1378605
18. Allouch, A., Koubâa, A., Abbes, T., Ammar, A.: Roadsense: smartphone application to estimate road conditions using accelerometer and gyroscope. IEEE Sens. J. **17**, 4231–4238 (2017). https://doi.org/10.1109/JSEN.2017.2702739
19. Mohan, P., Padmanabhan, V.N., Ramjee, R.: Nericell: rich monitoring of road and traffic conditions using mobile smartphones. In: Proceedings of the 6th ACM International Conference on Embedded Networked Sensor Systems, Raleigh, NC, USA, 5–7 November 2008, pp. 323–336. ACM, New York (2008). https://doi.org/10.1145/1460412.1460444

20. Yu, J., et al.: Senspeed: sensing driving conditions to estimate vehicle speed in urban environments. IEEE Trans. Mob. Comput. **15**, 202–216 (2016). https://doi.org/10.1109/TMC.2015.2411270
21. Chong, Y.S., Tay, Y.H.: Abnormal event detection in videos using spatiotemporal autoencoder. In: Cong, F., Leung, A., Wei, Q. (eds.) ISNN 2017. LNCS, vol. 10262, pp. 189–196. Springer, Cham (2017). https://doi.org/10.1007/978-3-319-59081-3_23
22. Xu, J., et al.: Stacked sparse autoencoder (SSAE) for nuclei detection on breast cancer histopathology images. IEEE Trans. Med. Imaging **35**(1), 119–130 (2015). https://doi.org/10.1109/TMI.2015.2458702
23. Almaslukh, B., AlMuhtadi, J., Artolim, A.: An effective deep autoencoder approach for online smartphone-based human activity recognition. Int. J. Comput. Sci. Netw. Secur **17**(4), 160–165 (2017)
24. Wang, A., Chen, G., Shang, C., Zhang, M., Liu, L.: Human activity recognition in a smart home environment with stacked denoising autoencoders. In: Song, S., Tong, Y. (eds.) WAIM 2016. LNCS, vol. 9998, pp. 29–40. Springer, Cham (2016). https://doi.org/10.1007/978-3-319-47121-1_3
25. Ducoffe, M., et al.: Anomaly detection on time series with wasserstein GAN applied to PHM. PHM Applications of Deep Learning and Emerging Analytics. Int. J. Prognostics Health Manag. Rev. (Special Issue) (2019).
26. Li, D., Chen, D., Jin, B., Shi, L., Goh, J., Ng, S.-K.: MAD-GAN: multivariate anomaly detection for time series data with generative adversarial networks. In: Tetko, I.V., Kůrková, V., Karpov, P., Theis, F. (eds.) ICANN 2019. LNCS, vol. 11730, pp. 703–716. Springer, Cham (2019). https://doi.org/10.1007/978-3-030-30490-4_56
27. Du, B., Zhang, L.: A discriminative metric learning based anomaly detection method. IEEE Trans. Geosci. Remote Sens. **52**(11), 6844–6857 (2014). https://doi.org/10.1109/TGRS.2014.2303895
28. Akbari, A., Jafari, R.: An autoencoder-based approach for recognizing null class in activities of daily living in-the-wild via wearable motion sensors. In: ICASSP 2019–2019 IEEE International Conference on Acoustics, Speech and Signal Processing (ICASSP). IEEE (2019). https://doi.org/10.1109/ICASSP.2019.8682161
29. Mora, N., et al.: Detection and analysis of heartbeats in seismocardiogram signals. Sensors **20**(6) (2020). https://doi.org/10.3390/s20061670
30. Park, D., Hoshi, Y., Kemp, C.C.: A multimodal anomaly detector for robot-assisted feeding using an lstm-based variational autoencoder. IEEE Robot. Autom. Lett. **3**(3), 1544–1551 (2018). https://doi.org/10.1109/LRA.2018.2801475
31. Kingma, D.P., et al.: Semi-supervised learning with deep generative models. In: Advances in Neural Information Processing Systems (2014). https://doi.org/10.5555/2969033.2969226

On the Legal Revision in PROLEG Program

Wachara Fungwacharakorn(✉), Kanae Tsushima, and Ken Satoh

National Institute of Informatics, Sokendai University, Tokyo, Japan
{wachara.f,k_tsushima,ksatoh}@nii.ac.jp

Abstract. PROLEG is a legal reasoning system that specializes in representing rules and exceptions separately. This system, as well as most of the computational legal reasoning systems, works based on the literal interpretation of legal rules. However, in some exceptional cases, the literal interpretation of legal rules may create unexpected consequences. In this paper, we introduce the idea of *responsible case assignment* for tracing back the intended precedents by lawmakers. Based on the responsible case assignment, we propose an algorithm for revising a generated PROLEG rulebase so that the revised rulebase covers the representative precedents intended by lawmakers as well as the new exceptional case.

1 Introduction

Researchers have long been interested in representing statutory laws by computers. One of those representations is PROLEG [4] developed to suit the ultimate fact theory [3] of the Japanese Civil Code used for the judgement process in civil code litigation. The civil code reasoning system by PROLEG has now around 2500 rules and exceptions which covers the most part of contract laws. A set of rules and exceptions has been compiled by graduates from the top law school in Japan and checked correctness by representing and solving all yes/no questions to use the ultimate fact theory in civil code bar examinations, which are based on codified factors in the Japanese Civil Code.

However, real cases may consist of new factors that were not considered before. Since PROLEG, as well as most of the legal systems, expresses legal reasoning based on codified factors, the system interprets the law in plain meaning without considering these new factors. Hence, it may cause unexpected consequences. In a judgement process, if unexpected consequences occur from interpretations, the court may describe the exceptional case with the new factor and revise the legal rule to cover the exceptional case. It requires much effort to guarantee this exhaustive consideration.

In this paper, we examine the idea to revise legal rules in a PROLEG program. In this preliminary step, a PROLEG program is presumed to be generated from a casebase. Intuitively, we trace back this casebase, which contains precedents intended by the lawmaker then we revise the legal rules to cover a new exceptional case as a new precedent.

K. Yada et al. (Eds.): JSAI 2020, AISC 1357, pp. 25–36, 2021.
https://doi.org/10.1007/978-3-030-73113-7_3

The paper is structured as follows. Sections 2 visualizes unexpected consequences in law using the example case. Section 3 describes inputs and outputs of Legal Revision, which are PROLEG programs [4] and a new exceptional judged case. Section 4 reports the details inside Legal Revision. The final section concludes and provides the future works.

2 Example Case

Here is an example case for illustrating the unexpected consequences in law.

> The plaintiff made a lease contract for his room between him and the defendant. When the defendant returned home for a while, he let his son use the room. Then, the plaintiff claimed that the contract was ended by his cancellation for the reason that the defendant subleases without permission.

The related piece of law in this case is Japanese Civil Code Article 612, which is stated as follows.

> **Phrase 1:** A Lessee may not assign the lessee's rights or sublease a leased thing without obtaining the approval of the lessor.
> **Phrase 2:** If the lessee allows any third party to make use of or take profits from a leased thing in violation of the provisions of the preceding paragraph, the lessor may cancel the contract.

From the literal interpretation of this article, the cancellation is valid. The third party in this case is the defendant's son and the defendant allowed his son to use the room without obtaining the approval of the lessor (the plaintiff in this case). However, it seems too harsh in this exceptional situation as the court decided as follows.

> Phrase 2 is not applicable in exceptional situations where the sublease does not harm the confidence between a lessee and a lessor, and therefore the lessor cannot cancel the contract unless they prove the lessee's destruction of confidence.

In this court decision, the court introduced the idea of destruction of confidence as an exception of Phrase 2 to prevent the unexpected consequence from the literal interpretation of Article 612. In this paper, we would like to formalize the process of legal revision to cover such a new exceptional case.

3 Overview and Preliminary Definitions

Figure 1 shows an overview of Legal Revision in PROLEG program. The inputs of Legal Revision are an original PROLEG program representing statutes and a new exceptional judged case. An output is a revised PROLEG program. In this section, we provide preliminary definitions of PROLEG programs and a new exceptional judged case.

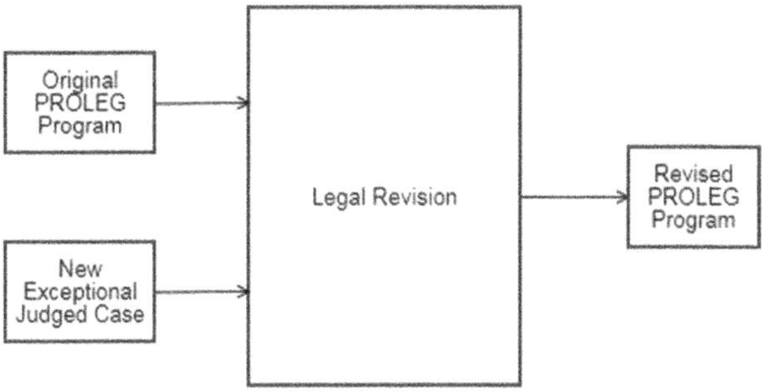

Fig. 1. A diagram showing inputs and an output of Legal Revision

3.1 PROLEG Program

In this paper, we use a PROLEG program to represent statutes. PROLEG programs occur in both an input and an output of Legal Revision. It was developed for representing the ultimate fact theory [3] of Japanese Civil Code. Although the ultimate fact theory was represented in PROLOG [5], it turns out that lawyers are not familiar with the semantics of negation as failure in PROLOG. So, PROLEG was invented to express familiar reasoning schemes by lawyers, namely, reasoning with general rules and exceptions expressed as follows.

Definition 1 (Rule and Exception). *We express a rule R in the form $h \Leftarrow b_1, \ldots, b_n$. where h, $b_1, \ldots, b_n$ are propositions. Sometimes, we express $h \Leftarrow B$ where $B = \{b_1, \ldots, b_n\}$. We denote $head(R) = h$ as a head of the rule and $body(R) = \{b_1, \ldots, b_n\}$ as a body of the rule. We express $h \Leftarrow .$ (called a fact rule) if the body of the rule is empty. We express an exception in the form $exception(h, e)$. where h and e are propositions. It expresses a rebuttal of h by e [4].*

A PROLEG rulebase (or simply called a rulebase is this paper) is a set of rules and exceptions that does not contain a fact rules and factors in a rulebase cannot occur in an exception. It reflects that an exception of legal rule should be rebuttable while facts cannot be rebutted. In other words, an exception of an exception is allowed but an exception of a fact is not allowed. A PROLEG rulebase is defined as follows.

Definition 2 (PROLEG rulebase). *Let S_r be a set of finite rules; S_e be a set of finite exceptions; and $RB = S_r \cup S_e$. We denote $rules(RB) = S_r$; $exceptions(RB) = S_e$; $pred(RB)$ as a set of all propositions that occur in RB; and, $factors(RB)$ as a set of all propositions that occur in RB but does not occur in a head of a rule. We call RB a rulebase when there are no fact rules $f \Leftarrow .$ in $rules(RB)$ and for every $exception(h, e) \in exceptions(RB)$, $h, e \notin factors(RB)$.*

Table 1 shows the representation of Japanese Civil Code Article 612 in PROLEG [4].

Table 1. PROLEG rulebase representing Japanese Civil Code Article 612

```
cancellation_due_to_sublease <=
    agreement_of_lease_contract, handover_to_lessee,
    agreement_of_sublease_contract, handover_to_sublessee,
    using_leased_thing, manifestation_cancellation.

exception(cancellation_due_to_sublease,approval_of_sublease).

approval_of_sublease <= approval_before_cancellation.
```

Let RB be a rulebase and FB denotes a factbase be a set of propositions such that $FB \cap pred(RB) \subseteq factors(RB)$. We call a pair $\langle FB, RB\rangle$ a legal database. The semantics of the legal database is called an extension in the same manner of the stable model of a normal logic program.

Definition 3 (Extension). *Let $\langle FB, RB\rangle$ be a legal database. The semantics of the legal database is called an extension obtained by.*

1. *Let RB be a rulebase and M be a set of propositions. We define a set of applicable rules w.r.t. M,RB^M as follows:*
   $\{R \in rules(RB) | \nexists exception(head(R), E) \in exceptions(RB) \text{ s.t. } E \in M\}$.
2. *Let M be a set of propositions and R be a rule. M* satisfies *R if this following condition is satisfied: if $body(R) \subseteq M$ then $head(R) \in M$.*
3. *A set of propositions M is an* extension *of a legal database $\langle FB, RB\rangle$ if $M = min(RB^M \cup \{f <= .|\forall f \in FB\})$ where $min(T)$ is the minimum model of a set of rules T (the minimum set in the sense of set inclusion such that M satisfies every rule in T).*
4. *Let p be a proposition. We denote $\langle FB, RB\rangle \vdash p$ if and only if p is in the extension of $\langle FB, RB\rangle$.*

For example, let RB be a rulebase in Table 1 and FB =

```
{agreement_of_lease_contract,handover_to_lessee,
agreement_of_sublease_contract,handover_to_sublessee,
using_leased_thing,manifestation_cancellation,
approval_before_cancellation}
```

Then, an extension of $\langle FB, RB\rangle$ is as follows, which represents the literal interpretation of the law.

```
{agreement_of_lease_contract,handover_to_lessee,
agreement_of_sublease_contract,handover_to_sublessee,
using_leased_thing,manifestation_cancellation,approval_of_sublease}
```

3.2 New Exceptional Judged Case

The second input of Legal Revision is a new exceptional judged case. When a new case leads to an unexpected consequence by the literal interpretation of the law, the court usually introduces a new factor that has not been codified in the law to the new case. We define a new exceptional judged case as follows.

Definition 4 (Judged Case). *Let RB be a rulebase and p be a non-factor proposition ($p \in pred(RB) \setminus factors(RB)$). We say a pair (C, J) is a judged case w.r.t. RB when C is a factbase w.r.t. RB and $J \in \{\text{'+'}, \text{'−'}\}$. We say C or (C, J) is new w.r.t. RB when $C \not\subseteq factors(RB)$ i.e. C contains extra factors that do not occur in RB. We say (C, J) is exceptional w.r.t. RB and p when $\langle C, RB\rangle \nvdash p$ but $J = \text{'+'}$; or $\langle C, RB\rangle \vdash p$ but $J = \text{'−'}$.*

For example, let RB be a rulebase in Table 1 and $(C, \text{'−'})$ be a judged case such that $C =$

```
{agreement_of_lease_contract,handover_to_lessee,
agreement_of_sublease_contract,handover_to_sublessee,
using_leased_thing,manifestation_cancellation,parent_sublease}
```

C is a new case since it contains an extra fact `parent_sublease` $\notin$ $factors(RB)$. and $(C, \text{'−'})$ is said to be an exceptional judged case w.r.t RB and p = `cancell-ation_due_to_sublease` since $\langle C, RB\rangle \vdash p$. This judged case actually represents an example case with p as an unexpected consequence.

4 Legal Revision

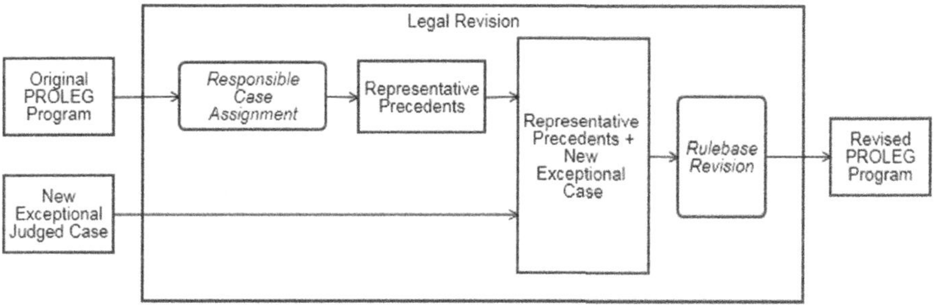

Fig. 2. A diagram showing details inside Legal Revision

Figure 2 shows details inside Legal Revision in PROLEG program. Firstly, the system translates an original PROLEG program into *representative precedents*. Then, the system merges a new exception judged case into representative precedents. After that, the system translates the new representative precedents back to a revised rulebase. The details of each part are described as follows.

4.1 Representative Precedents

Representative precedents are judged cases that reflect the lawmaker intention. We use an abstract argumentation for case-based reasoning (AA-CBR) [2] as the reasoning behind precedents. Given a set of judged cases CB called a casebase with a default judged case $(\{\}, J_0)$, an argumentation framework is formed to predict a judgement as follows. (We denote the complement of J by $\bar{J}$, namely $\bar{J} = \text{'+'}$ if $J = \text{'−'}$; and $\bar{J} = \text{'−'}$ if $J = \text{'+'}$.)

Definition 5 (Super-nearest Counter-judged Case (s.n.c.)). *Let CB be a casebase and $(C, J) \in CB$. A judged case $(N, \bar{J})$ is a super-nearest counter-judged case (s.n.c. in shorts, not always unique) to (C, J) w.r.t. CB when $C \subsetneq N$, and no other $(N', \bar{J}) \in CB$ such that $C \subsetneq N' \subsetneq N$.*

Definition 6 (AF corresponding to a casebase). *Let CB be a casebase. $AF = (CB, attacks)$ is an argumentation framework (AF) corresponding to CB where attacks is a set of $(X, J) \rightsquigarrow (Y, \bar{J})$ for all pairs $(X, J), (Y, \bar{J}) \in CB$ such that $(X, \bar{J})$ is a s.n.c. to (Y, J) w.r.t. CB.*

Definition 7 (Predicted Judgement in AA-CBR). *To predict a judgement for a case N, let $CB_N = \{(C, J) \in CB | C \subseteq N\}$. A predicted judgement of N is J_0 if $(\{\}, J_0)$ is in the ground extension of the corresponding argumentation framework of CB_N. Otherwise, a predicted judgement of N is $\bar{J}_0$* [1,2].

Example 1. Let $CB_0 = \{(\{\}, \text{'}-\text{'}), (\{a\}, \text{'}+\text{'}), (\{a, b\}, \text{'}-\text{'}), (\{c\}, \text{'}+\text{'})\}$ and $N = \{b, c, d\}$. We get that $CB_N = \{(\{\}, \text{'}-\text{'}), (\{c\}, \text{'}+\text{'})\}$ and the ground extension of the corresponding argumentation framework of CB_N is $G = \{(\{c\}, \text{'}+\text{'})\}$. Hence, the predicted judgement of N is '+' since $(\{\}, \text{'}-\text{'}) \notin G$.

From the definition, determining predicted judgement requires only judged cases that are transversely connected to a default judged case $(\{\}, J_0)$ via *attacks*. This is corresponding to the minimization of casebase representation using critical cases [1,5] which is defined as follows.

Definition 8 (Critical casebase). *Let CB be a casebase and $(CB, attacks)$ be an argumentation framework corresponding to CB. A judged case $(C, J) \in CB$ is critical if and only if (C, J) is a default judged case $(C = \{\})$ or $(C, J) \rightsquigarrow (B, \bar{J}) \in attacks$ where $(B, \bar{J})$ is critical. A critical casebase means a casebase in which all judged cases are critical.*

Previous works [1,5] suggest quite similar ideas to generate a rulebase from a critical casebase with $(\{\}, \text{'}-\text{'})$ as follows.

Definition 9 (Generated rulebase). *Let CB be a critical casebase with $(\{\}, \text{'}-\text{'})$, $(CB, attacks)$ be an argumentation framework corresponding to CB, p_0 be a proposition, and $p_{C,D}$ be a reified proposition of cases C and D and distinct to p_0. A rulebase RB is generated from a critical casebase CB as follows.*

$$RB = \{p_0 \Leftarrow p_{C,\{\}} | (C, \text{'}+\text{'}) \rightsquigarrow (\{\}, \text{'}-\text{'}) \in attacks\} \cup$$
$$\{p_{C,D} \Leftarrow C \setminus D | (C, J) \rightsquigarrow (D, \bar{J}) \in attacks\} \cup$$
$$\{exception(p_{C,D}, p_{B,C}) | (B, J) \rightsquigarrow (C, \bar{J}) \in attacks, (C, \bar{J}) \rightsquigarrow (D, J) \in attacks\}.$$

Example 2. For example, $CB_1 = \{(\{\}, \text{'}-\text{'}), (\{a\}, \text{'}+\text{'}), (\{a, b\}, \text{'}-\text{'}), (\{c\}, \text{'}+\text{'})\}$ is a critical casebase. RB_0 is a generated rulebase from CB_1.

$$RB_0 = \left\{ \begin{array}{ll} p_0 \Leftarrow r1. & p_0 \Leftarrow r2. \\ r1 \Leftarrow a. & r2 \Leftarrow c. \\ exception(r1, e1). & \\ e1 \Leftarrow b. & \end{array} \right\}$$

where $r1$ is a rectified proposition of $\{a\}$ and $\{\}$; $e1$ is a rectified proposition of $\{a, b\}$ and $\{a\}$; and $r2$ is a rectified proposition of $\{c\}$ and $\{\}$.

4.2 Responsible Case Assignment

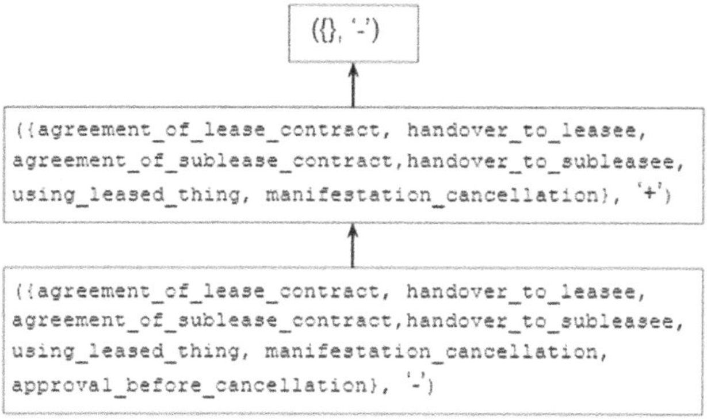

Fig. 3. Representative precedents of a rulebase in Table 1

Since we can generate a rulebase from a critical casebase, in the other way round, we can trace back the originate critical casebase from a given generated rulebase. To trace back the critical casebase, we introduce a responsible case assignment K. Intuitively, a responsible case assignment is an invert of reified proposition $p_{C,D}$ of cases C and D in Definition 9, namely for every reified propositions $p_{C,D}$, $(p_{C,D}, C, D, J)$ is in K.

Definition 10 (Responsible case assignment). *Let RB be a generated rulebase with a top proposition* p_0. *K denotes a* responsible case assignment *w.r.t. RB, which is a subset of* $(pred(RB) \setminus (\{p_0\} \cup factors(RB)) \times P(factors(RB)) \times \{\text{'+'}, -\}$ *such that*

- $(q, C, \{\}, \text{'+'}) \in K$ *for all* $(p_0 \Leftarrow q.), (q \Leftarrow C.) \in RB$
- $(q, B \cup C, C, \bar{J}) \in K$ *for all* $exception(p, q) \in RB, (p, C, D, J) \in K, (q \Leftarrow B.) \in RB$

Example 3. Let RB_1 be a rulebase as follows. K_1 is a responsible case assignment w.r.t. RB_1.

$$RB_1 = \left\{ \begin{array}{ll} p_0 \Leftarrow r1. & p_0 \Leftarrow r2. \\ r1 \Leftarrow a. & r2 \Leftarrow c. \\ exception(r1, e1). & \\ e1 \Leftarrow b. & \end{array} \right\} \quad K_1 = \left\{ \begin{array}{l} (r1, \{a\}, \{\}, \text{'+'}), \\ (r2, \{c\}, \{\}, \text{'+'}), \\ (e1, \{a, b\}, \{a\}, \text{'}-\text{'}) \end{array} \right\}$$

Theorem 1. *Let CB be a critical casebase with $(\{\}, \text{'}-\text{'})$; RB be a rulebase generated from CB with a top proposition p_0; and K be a responsible case assignment w.r.t. RB. $CB = \{(\{\}, \text{'}-\text{'})\} \cup \{(C, J) | (p, C, D, J) \in K\}$.*

Proof. Let $CB' = \{(\{\}, \text{'}-\text{'})\} \cup \{(C, J) | (p, C, D, J) \in K\}$ with a corresponding argumentation framework $(CB', attacks')$. Then, we get that the rulebase generated from CB' is actually a variant of RB such that p is a rectified proposition of C and D for all $(p, C, D, J) \in K$. Hence, $CB = CB'$.

Such CB in Theorem 1 is a casebase of *representative precedents* of the rulebase RB. For example, representative precedents of a rulebase in Table 1 is shown in Fig. 3. We get that the time complexity of responsible case assignment is $O(k\|rules(RB)\|)$ where k is the maximum depth, namely the maximum number of exceptions passing from the top proposition p_0 to each proposition in $pred(RB)$.

4.3 Rulebase Revision

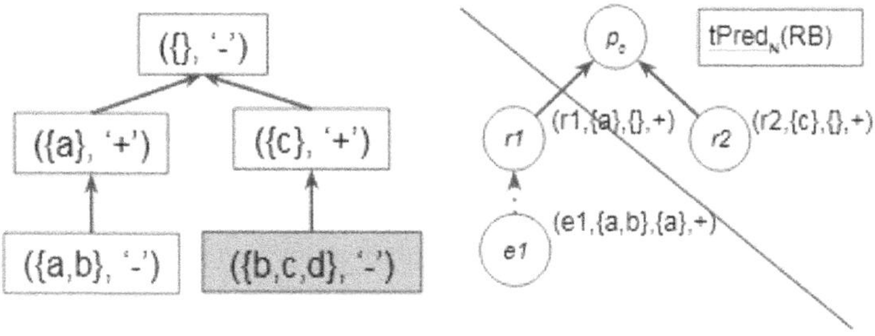

Fig. 4. A casebase when including a new exceptional judged case (left) and trimming (right)

Rulebase revision aims to revise the original rulebase so that it covers a new exceptional judged case. From Example 1, $(\{b, c, d\}, \text{'}-\text{'})$ is a new exceptional judged case w.r.t. RB_1. When we add $(\{b, c, d\}, \text{'}-\text{'})$ to the casebase, the argumentation framework corresponding to the new casebase is modified by adding attacks only those are related to the new exceptional case as in Fig. 4 (left). Hence, we can modify the rulebase by append a rulebase RB_a such that if RB is an original rulebase, $RB \cup RB_a$ becomes a revised rulebase from RB that covers the new exceptional case.

Moreover, it is not necessary to know all representative precedents to revise a rulebase. It requires only judged cases of which the new exceptional judged case is a super-nearest counter-judged case. To reduce the generation of representative precedents, we use the idea of trimming as follows.

Definition 11 (Trimming). *Let CB be a critical casebase with $(\{\}, `-')$; RB be a rulebase generated from CB with a top proposition p_0; K be a responsible case assignment w.r.t. RB; and N be a new case w.r.t. RB. We denote a set of trimming propositions by* $tPred_N(RB) = \{p_0\} \cup \{p \in pred(RB) | (p, C, D, J) \in K, C \subseteq N\}$.

From Example 3, let $N = \{b, c, d\}$, we get that $tPred_N(RB) = \{p_0, r2\}$ as in Fig. 4 (right).

Theorem 2. *Let CB be a critical casebase with $(\{\}, `-')$; RB be a rulebase generated from CB with a top proposition p_0; K be a responsible case assignment w.r.t. RB; and (N, J) be a new exceptional judged case w.r.t. RB w.r.t. p_0. There is a non-empty case C such that (N, J) is a s.n.c. of $(C, \bar{J})$ w.r.t. CB if and only if there is $(p, C, D, \bar{J}) \in K$ such that $p \in tPred_N(RB)$ and no $exception(p, q) \in RB$ such that $q \in tPred_N(RB)$.*

Proof. If there is a non-empty case C such that (N, J) is a s.n.c. of $(C, \bar{J})$ w.r.t. CB then there is $(p, C, D, \bar{J}) \in K$. Suppose there is $exception(p, q) \in RB$ such that $q \in tPred_N(RB)$. Then, there is a non-empty case B such that $(q, B \cup C, C, J) \in K$. Hence, $(B \cup C, J) \in CB$ and $C \subsetneq B \cup C \subsetneq N$. Thus, (N, J) is not a s.n.c. of $(C, \bar{J})$ w.r.t. CB. This leads to contradiction.

Suppose there is $(p, C, D, \bar{J}) \in K$ such that $p \in tPred_N(RB)$ and there is a case N' such that $C \subsetneq N' \subsetneq N$. Then, there is $q \in tPred_N(RB)$ such that $exception(p, q) \in RB$. This leads to contradiction.

From Theorem 2, we design Algorithm 1 for revising a generated rulebase using the responsible case assignment for covering a new exception judged case and trimming propositions to reduce the generation of the representative precedents. We defined the revised rulebase as in Definition 12.

Definition 12 (Revised rulebase). *Let CB be a critical casebase with $(\{\}, `-')$; RB be a rulebase generated from CB with a top proposition p_0; K be a responsible case assignment w.r.t. RB; and (N, J) be a new exceptional judged case w.r.t. RB. We say RB' is a rulebase revised from RB w.r.t. (N, J) defined as follows.*

(i) if no $(C, `+') \in CB$ that $C \subseteq N$ then $RB' = RB \cup \{p_0 \Leftarrow p_{N,\{\}}. p_{N,\{\}} \Leftarrow N.\}$. where $p_{N,\{\}}$ is a new rectified proposition.

(ii) otherwise
- *Let $\Gamma = \{R \in rules(RB) | head(R) \in tPred_N(RB)$ and R is an applicable rule w.r.t. $tPred_N(RB)\}$.*
- *$RB' = RB \cup \{exception(head(R), e_R). e_R \Leftarrow N \setminus C | R \in \Gamma, (head(R), C, D, \bar{J}) \in K\}$*

where e_R is a new rectified proposition w.r.t. R.

Example 4. RB'_1 is a revised rulebase from RB_1 in Example 3 w.r.t. $(\{b, c, d\}, \text{'}-\text{'})$ and K'_1 is a responsible case assignment w.r.t. RB'_1.

$$RB'_1 = \left\{ \begin{array}{ll} p_0 \Leftarrow r1. & p_0 \Leftarrow r2. \\ r1 \Leftarrow a. & r2 \Leftarrow c. \\ exception(r1, e1). & exception(r2, e2). \\ e1 \Leftarrow b. & e2 \Leftarrow b, d. \end{array} \right\}$$

$$K'_1 = \left\{ \begin{array}{ll} (r1, \{a\}, \{\}, \text{'+'}), & (r2, \{c\}, \{\}, \text{'+'}), \\ (e1, \{a, b\}, \{a\}, \text{'}-\text{'}) & (e2, \{b, c, d\}, \{c\}, \text{'}-\text{'}) \end{array} \right\}$$

Theorem 3. *From Definition 12, RB' is a rulebase generated from $CB \cup \{(N, J)\}$.*

Proof. Since RB' in (i) is obvious, we only prove (ii). Let K and K' be responsible case assignments w.r.t. RB and RB' respectively; $(CB, attacks)$ and $(CB \cup \{(N, J)\}, attacks')$ be argumentation frameworks corresponding to CB and $CB \cup \{(N, J)\}$ respectively. We get that $K' = K \cup \{(e_R, N, C, J) | C \in C_m\}$ and Γ is a set of all rules R that no $exception(head(R), q) \in CB$ such that $q \in tPred_N(RB)$ Hence, (N, J) is a s.n.c. of $(C, \bar{J})$ w.r.t. CB such that $(head(R), C, D, \bar{J}) \in K\}$. All elements in $attacks' \setminus attack$ are in the form $(N, J) \rightsquigarrow (C, \bar{J})$. Hence, RB' is a rulebase generated from $CB \cup \{(N, J)\}$.

For another example, let RB be a rulebase in Table 1 and the new exceptional judged case $(C, \text{'}-\text{'})$ such that $C =$

`{agreement_of_lease_contract,handover_to_lessee,`
`agreement_of_sublease_contract,handover_to_sublessee,`
`using_leased_thing,manifestation_cancellation,parent_sublease}`

The result is a revised rulebase as in Table 2. It tells that `cancellation_due_to_sublease` has a new exception derived from `parent_sublease`, which is an extra fact and a difference of the new case and the related representative precedents (Fig. 3). Since we use trimming to reduce the execution of responsible case assignment, the time complexity of rulebase revision is $O(k\|tPred_N(RB)\|)$ where k is the maximum depth, namely the maximum number of exceptions passing from the top proposition p_0 to each proposition in $\|tPred_N(RB)\|$.

Table 2. A revised rulebase corresponding to Section 3.2

```
cancellation_due_to_sublease <=
    agreement_of_lease_contract, handover_to_lessee,
    agreement_of_sublease_contract, handover_to_sublessee,
    using_leased_thing, manifestation_cancellation.

exception(cancellation_due_to_sublease,approval_of_sublease).

approval_of_sublease <= approval_before_cancellation.

exception(cancellation_due_to_sublease,new_exception).

new_exception <= parent_sublease.
```

Algorithm 1. Revising a generated rulebase for covering a new exceptional judged case

Given a generated rulebase RB **Input** a new exceptional judged case (N, J)
Output an amendment rulebase RB_a such that $RB \cup RB_a$ becomes a revised rulebase from RB w.r.t. (N, J)

procedure REVISE(N,J)
  Let $G =$ GAMMA(p_0,N)
  if $G = \{p_0\}$ **then** return $\{p_0 \Leftarrow p_N . p_N \Leftarrow N.\}$.
  else
    Let $RB_a = \{\}$
    for all $p \in G$ **do**
      Let $(C_p, J_p) =$ RESP-CJ(p)
      if $J_p = \bar{J}$ **then**
        $RB_a = RB_a \cup \{exception(p, e_p).\} \cup \{e_p \Leftarrow N \setminus C_p\}$
    return RB_a
procedure GAMMA(p,N)
  if $p = p_0$ **then**
    Find $p_0 \Leftarrow q$ s.t. $q \Leftarrow C_q$ and $C_q \subseteq N$
    if there is no such q **then** return $\{p_0\}$
    else
      Let $G = \{\}$
      for all $p_0 \Leftarrow q$ s.t. $q \Leftarrow C_q$ and $C_q \subseteq N$ **do**
        $G = G \cup$ GAMMA(q,N)
      return G
  else
    Find $exception(p, q)$ s.t. RESP-CJ(q) $= (C_q, J_q)$ and $C_q \subseteq N$
    if there is no such q **then** return $\{p\}$
    else
      Let $G = \{\}$
      for all $exception(p, q)$ s.t. RESP-CJ(q) $= (C_q, J_q)$ and $C_q \subseteq N$ **do**
        $G = G \cup$ GAMMA(q,N)
      return G
procedure RESP-CJ(q)
  Find $p_0 \Leftarrow q$ s.t. $q \Leftarrow C_q$ and $C_q \subseteq N$
  if there is $p_0 \Leftarrow q$ **then**
    Find $q \Leftarrow C_q$ return $(C_q, \text{'+'})$
  else
    Find $exception(p, q)$ and $q \Leftarrow C_q$
    Let $(C_p, J_p) =$ RESP-CJ(p) return $(C_p \cup C_q, \bar{J_p})$

5 Conclusion and Future Works

This paper describes the idea to trace back cases that the lawmaker intended using a *responsible case assignment* corresponding to the generation of a rulebase from a casebase as in [1]. Based on a responsible case assignment, we proposed the algorithm revising the generated rulebase. The revised rulebase includes the new exceptional judged case, which the court usually introduces new facts that have never occurred in the statute. We also introduce trimming in the rulebase revision to reduce the execution of responsible case assignment. This work still limits on the generated rulebase and only implies the intention of the lawmaker from the converting critical casebase. Therefore, in future, we would like to extend the revising algorithm for other specifications of rulebase or other implications of the intention of the lawmaker.

Acknowledgement. This work was supported by JSPS KAKENHI Grant Number 17H06103.

References

1. Athakravi, D., Satoh, K., Law, M., Broda, K., Russo, A.: Automated inference of rules with exception from past legal cases using ASP. In: International Conference on Logic Programming and Nonmonotonic Reasoning, pp. 83–96. Springer (2015)
2. Cyras, K., Satoh, K., Toni, F.: Abstract argumentation for case-based reasoning. In: Fifteenth International Conference on the Principles of Knowledge Representation and Reasoning (2016)
3. Ito, S.: Basis of Ultimate Facts. Yuhikaku (2001)
4. Satoh, K., et al.: PROLEG: an implementation of the presupposed ultimate fact theory of Japanese civil code by PROLOG technology. In: Onada, T., Bekki, D., McCready, E. (eds.) New Frontiers in Artificial Intelligence. Lecture Notes in Computer Science, pp. 153–164. Springer, Heidelberg (2011)
5. Satoh, K., Kubota, M., Nishigai, Y., Takano, C.: Translating the Japanese presupposed ultimate fact theory into logic programming. In: Proceedings of the 2009 Conference on Legal Knowledge and Information Systems: JURIX 2009: The Twenty-Second Annual Conference, pp. 162–171. IOS Press, Amsterdam (2009)

Viewpoint Planning Based on Uncertainty Maps Created from the Generative Query Network

Kelvin Lukman[1(✉)], Hiroki Mori[2], and Tetsuya Ogata[1,2,3]

[1] The Department of Intermedia Art and Science, Waseda University, Shinjuku City, Japan
kelvin-raku@moegi.waseda.jp, ogata@waseda.jp
[2] Future Robotics Organization, Waseda University, Shinjuku City, Japan
[3] National Institute of Advanced Industrial Science and Technology, Tokyo, Japan

Abstract. Most current research on robotic tasks such as grasping uses the single fixed viewpoint positioned to oversee the environment from above. However, there are cases when a single fixed viewpoint does not provide sufficient information to perform a designated robot task. Environments with high occlusion, or cases where objects have fragile components, it is difficult for robots to identify the task relevant viewpoint to achieve a task or goal. Herein, to reconstruct the world model with less effort, we propose a viewpoint planner method that uses uncertainty in scene representation from Generative Query Network(GQN) With this scene representation, we create an uncertainty map by calculating the pixel-wise variance of multiple predicted images for each query viewpoint. The results indicate that our solution is capable of taking a suitable view of an untrained object, therefore reducing the uncertainty. A suitable viewpoint is defined as one that improves prediction certainty by focusing an area in which the learning environment allows the highest uncertainty value. In a future study, we would like to propose a task specific viewpoint planner models for robotic tasks such as grasping objects with occlusion by extending this research.

Keywords: Generative Query Network · Uncertainty map · Viewpoint planning · Robotics · Manipulation

1 Introduction

Understanding the structure of the physical environment rather than the structure of a one-shot image from a single viewpoint is critical for intelligent systems. Most current research in robotic tasks such as grasping employs a single fixed viewpoint which is positioned to oversee the learning environment from above. However, there are situations where a single fixed viewpoint does not provide sufficient information to perform a designated robotic task. In high occlusion,

K. Yada et al. (Eds.): JSAI 2020, AISC 1357, pp. 37–48, 2021.
https://doi.org/10.1007/978-3-030-73113-7_4

environments or cases where objects have fragile components, identifying the task relevant viewpoint to achieve a task is difficult for robots.

A substantial amount of research on robot grasping adopts on data-driven approaches that rely on human demonstrations or 3D models of objects [2]. However, evaluation of these methods has shown that they do not cope well with challenges that arise in unstructured environments.

Applying human labels in RGB-D data to identify grasping regions [7,12] or point clouds to enable grasp detection [6] is a popular approach in data-driven learning frameworks for robotic grasping tasks. Grasping research which applies a database of synthetic point clouds annotated with grasp labels [9–11] has demonstrated remarkable precision in the grasping of everyday objects; however, creating and annotating 3D models of every possible object that the network might encounter would be difficult and inefficient.

Data-driven approaches that only rely on visual perception, such that employed in [8], employ data on prior grasping attempts to train models that are capable of predicting the probability of a successful subsequent grasp.

Most recent work on neural network application to robotic grasping tasks entails a fixed single camera that observes the learning environment and in most cases assumes that the robot will grasp objects from above. Data-driven approaches require human labels that may be less laborious when data is collected by robots [8,9]; however, these methods are time consuming and inefficient if we try to include every possible existing object.

In these cases the camera is always fixed above the learning environment, and it is assumed that the robot arm will always grasp objects from above. However, there are cases in which this setup may not provide sufficient information for the robot to successfully perform the designated task. Specifically, high occlusion environments will pose challenges to observation setups that rely on a single fixed camera. In these instances, the robot is unable to recognize the true state (position and orientation) of the object because of occlusions resulting from the single fixed viewpoint.

To resolve the challenges that accompany fixed single viewpoints in robot tasks, we propose a viewpoint planning method that utilizes scene representation networks (SRNs). As mentioned above, creating models and annotations of every possible object is time consuming and inefficient. Therefore, as SRNs are capable of representing 3D features in a world model using only 2D visual inputs, we selected this architecture for our research purpose.

This research was inspired by generative "world model" research, especially the capabilities of the Generative Query Network (GQN) [4] in the modelling of uncertainty with variability in its predictions. We conjectured that the model could contribute not only to prediction but also application of active perception in robotic tasks to understand the structure of the world via an internal representation.

Most robots are fitted with multiple and various sensors to allow them to perceive and navigate task environments, and the computational burden can be difficult to deal with. Therefore it is preferable to only infer the required information from visual perception.

We specifically apply the GQN because it is capable of modeling uncertainty through variability in its predictions. An uncertainty map is generated from the scene representation by calculating the pixel-wise variance of multiple predictions, which allows the neural network to identify areas of high uncertainty in the learning environment. The next most optimal camera viewpoint is determined by finding the viewpoint that incorporates the most uncertainty. A viewpoint with high uncertainty enables the robot to gain more and necessary information by observing that viewpoint and thus to perform its designated tasks.

Few works apply SRNs to viewpoint planning, and recent approaches primarily rely on reinforcement learning to determine the optimal viewpoint [14,16]. Our approach employs the uncertainty embodied in the world model of the neural network to identify the next suitable viewpoint. This approach allows human defined rules to be avoided, as well as repetitive exploration and long training times, which are common disadvantages of employing a reinforcement learning based method.

We propose a novel viewpoint planning method with two main components: (1) generating uncertainty maps by quantifying uncertainty in the world model of the neural network and (2) determining the next optimal viewpoint by finding a viewpoint that incorporates the highest value of uncertainty. An uncertainty map is generated from the scene representation by calculating the pixel-wise variance of multiple predictions, which allows the neural network to identify areas of high uncertainty in the learning environment, and is used to guide the camera viewpoint to obtain more information about the learning environment. An overview of the model is shown in Fig. 1.

In industrial environments, everything is organized and designed to help robots perform efficiently; therefore viewpoint planning may not be a necessity in these cases. However, we live in disorganized environments, and for robots to work and adapt to these environment, viewpoint planning is an essential capability. The solution proposed in this research may enable robots to deal with the perception bottlenecks that can arise when working in human society.

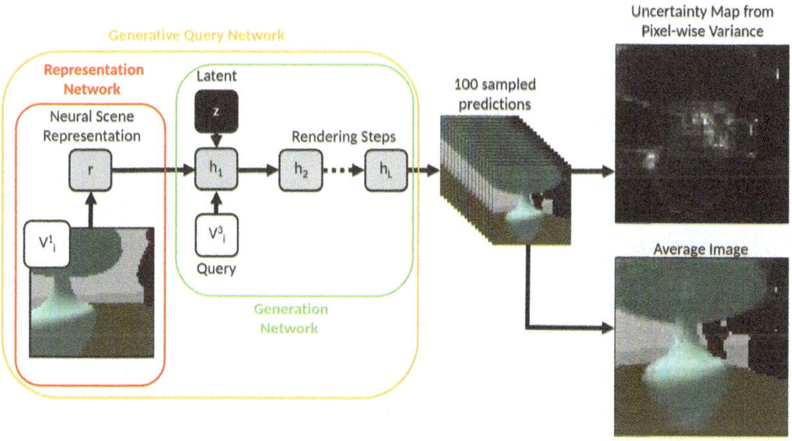

Fig. 1. Overview of the model.

2 Related Works

We propose a method for viewpoint optimization by applying uncertainty maps created from an SRN. SRNs are capable of representing and reconstructing 3D features from 2D inputs, and we apply this to an active vision inspired system capable of adaptively determining a suitable viewpoint for a robotic task. The primary fields relating to this research include SRN, active vision, and viewpoint planning.

2.1 Active Vision

Our proposed method is inspired by active vision, and intends to improve on viewpoint planning by selecting a viewpoint that can potentially increase information to construct a better understanding of surrounding objects for robots working in unfamiliar and unstructured environments. The field of active vision is defined as the study of modeling and control strategies for perception [1]. As humans, we perceive the world not in a passive manner, but in an active one. When we look, the pupils dilate or contract based on illumination, and we may turn or incline our heads to better observe an object within or outside our immediate field of view.

A survey of research on active vision applied to robotic systems [3] describe previous works that relied on sensors and 3D modeling to allow robots to understand their environments. Most of the described methods rely on novel exploration algorithms to investigate uncertainties in the environment.

We aim to simplify the formulation of rules for investigation by employing uncertainties present in the world model of neural networks. By identifying the uncertainties in the learning environment, the robot will be capable of exhibiting reactive behavior by moving its visual sensor to a more advantageous viewpoint to gain more information about the environment.

2.2 Scene Representation Networks

Current research on SRNs focuses on deriving 3D features from 2D images to generate geometrically accurate images from unknown viewpoints on the basis of at least a single 2D reference image [4,13,17,18].

As described in [17], the SRN follows a novel method to represent image formation as a differentiable ray-marching algorithm that generates continuous 3D structural representations from 2D images. It is demonstrated that the 3D representations are more distinguishable than predictions from a deterministic GQN. In [18], the rendered quality of predictions generated by the GQN is improved by implementing Epipolar Cross Attention (ECA) that adds geometrical information to the scene representation vector. Another work in scene representation [13] implements a generative adversarial network based method that learns explicit 3D features to disentangle 3D poses and identities to generate realistic novel views from natural images. Although SRN was applied in our method, our goal is quite different from those of other SRNs.

Most of the aforementioned SRN models are deterministic and aim to improve the accuracy of generated predictions, with the exception of GQN whose architecture is capable of modeling uncertainty in its predictions. Rather than developing a more deterministic model to generate high quality predictions, our approach requires the variability of predictions made with the GQN that is capable of representing uncertainty in its world representation and we use this to create an uncertainty map of the observed environment.

2.3 Viewpoint Planning

In [16], Sather and Zhang developed a viewpoint planner with deep reinforcement learning for autonomous harvesting. They used Deep Deterministic Policy Gradients method which provide reward feedback based on data obtained from a pretrained object detector. Our goal is very similar, but we aim to apply our method not only to autonomous harvesting or robot grasping, but also to other general applications. In contrast to their approach, which draws on deep reinforcement learning, we create an uncertainty map from the scene representation of the environment to guide the robot viewpoint.

A viewpoint selection method proposed in [14] applies reinforcement learning to answer questions based on what can be observed from the current viewpoint. During training, the model is rewarded every time it predicts the correct answer to the question provided. The system proposed in [14] will have difficulty in coping with cases in which the observed object does not have a known color or shape. Our viewpoint planning method does not require prior knowledge of existing shapes and colors and is therefore more applicable to various tasks.

3 Proposed Method

The proposed viewpoint planner utilizes a hill climbing algorithm to find the viewpoint that corresponds to the highest value of average variance of prediction by GQN. The system is illustrated in Fig. 2.

The model obtains at least one image and viewpoint information v_i^1 from the camera attached to the robot arm, which then creates a latent vector representation of the observed scene. The model generates predictions from multiple possible query viewpoints (unobserved) represented by $V_q^1, V_q^2, ...V_q^n$, and an uncertainty image is generated by calculating the pixel-wise variance of 100 predictions corresponding to a single query viewpoint. The accumulation of uncertainty images from multiple query viewpoints creates the uncertainty map of the environment, based on the given observations.

We define uncertainty as the pixel-wise variance calculated from 100 sampled predictions of a single query viewpoint, which we generate from the scene representation of the GQN. The sampling of prediction is generated by Gaussian distributions with the calculated latent variables z with μ and σ^2. The average variance is calculated for each uncertainty image, and the uncertainty image with the highest average variance is then selected using the hill climbing algorithm. From the selected uncertainty image, we obtain the corresponding query viewpoint, which then becomes the next optimal viewpoint. A robot arm motion is then generated to move the attached camera to the new viewpoint.

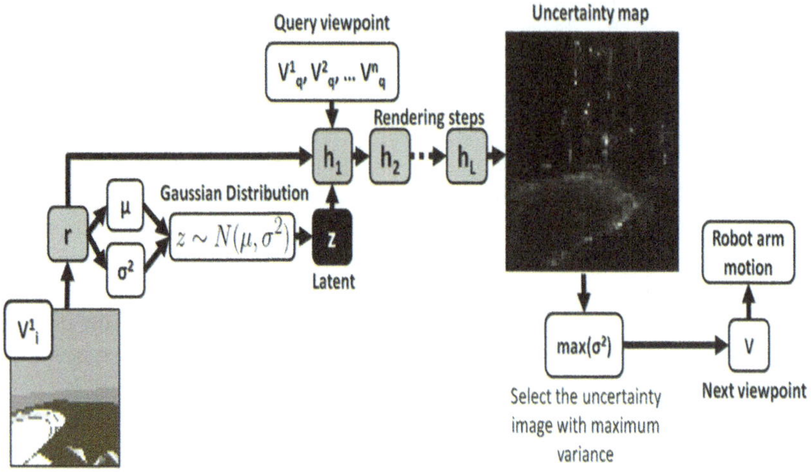

Fig. 2. Hill climbing algorithm applied to obtain viewpoint which observes highest value of average variance.

4 Experiments

4.1 Preliminary Experiment

Experiments were conducted to ascertain whether the quantitative measurement of uncertainty in the variability of predictions in GQN matches the qualitative change in the rendered predictions. We used the Shepard–Metzler 5-part dataset from the deepmind/gqn datasets[1] repository to evaluate the uncertainty metric, as it is easier to identify qualitative changes in the rendered predictions due to the simple nature of the object shapes.

Initially we contemplated the possibility of generating an uncertainty representation from only the latent vectors, which is the reason for the comparison of the relation between the variance of the latent vectors and the pixel-wise variance of the generated predictions. However, we found that the variance in the latent scene representation by GQN has a relatively opposing trend in relation to the variance of the rendered predictions as shown in Fig. 3. This was confirmed to be due to the architecture in DRAW[5], which is also implemented in GQN; therefore the variance changes because the latent representation undergoes several rendering steps before generating an output. Therefore, we chose to use the pixel-wise variance of predictions to create uncertainty maps.

[1] https://github.com/deepmind/gqn-datasets.

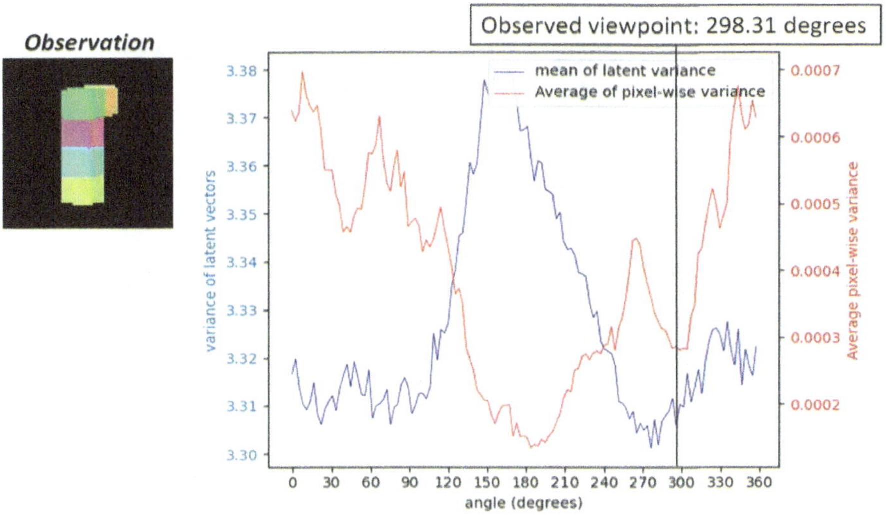

Fig. 3. Variance in the latent vector and the pixel-wise variance of 100 predictions, in relation to the change in viewpoint angle. The latent vector is 3-dimensional; therefore we compare the mean of the variance.

In Fig. 3, the observed viewpoint is a point sitting on the circumference of a 0.2 m-radius circle, and the x-axis shows the rotation angle of the viewpoint moving along the circumference in degrees. The subsequent angles before and after the observed viewpoint on the graph are unobserved viewpoints. A Shepard–Metzler object is positioned in the center of the 0.2 m radius circle. The pixel-wise variance of 100 sampled predictions for each unobserved viewpoint is generated from the latent scene representation of one observed viewpoint, represented by the red line in Fig. 3. For this particular observed viewpoint of the Shepard–Metzler object, the variance of the latent scene representation is highest when the GQN model attempts to predict the image observed from a query viewpoint on the opposing side of the observed view point (between 0–30°).

4.2 Robotic Application

For our purpose, the learning environment is built using the Gazebo simulator. We developed the environment with Rethink Robotics' Baxter Robot as the main subject. This is primarily because we intend to extend this work onto dual-armed robots capable of optimizing its viewpoint with one arm and performing the designated task with the other arm.

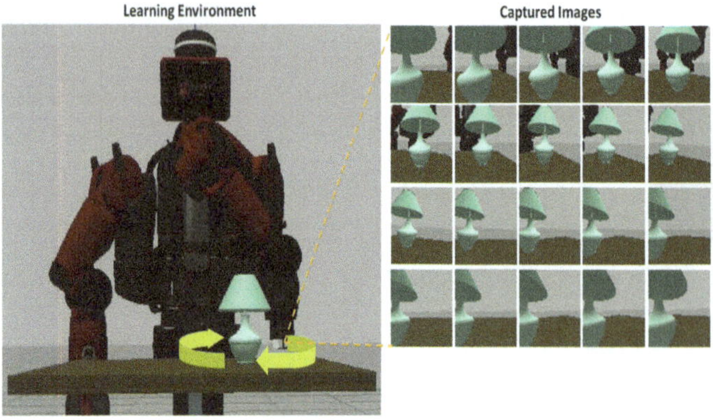

Fig. 4. Learning environment used in the Gazebo simulator.

As shown in Fig. 4, the simulated environment consists of Baxter and a table positioned directly in front of it. A variety of objects are spawned individually on top of the table. These objects were obtained from an open source dataset of 3D Gazebo models called 3DGEMS, used in [15], which investigates the effect of color spaces on the detectability of colored objects. From this large dataset, we selected the categories Decoration, Earthquake, Electronics, Food, Kitchen, and Stationery, based on the size of objects within these categories. The objects used in the learning environment are small enough to fit in a dome with a 0.2-m radius.

Although the Baxter robot is already equipped with two cameras on either arm, the original orientations of the cameras limit the number of viewpoints that can be reached by the arm due to its limited flexibility. Reaching viewpoints positioned behind the object with the default Baxter arm cameras was especially difficult. As an alternative, the left electric gripper of the Baxter robot was replaced with a simple camera oriented perpendicular to the z-axis of the Baxter wrist. This configuration enabled us to incorporate more of the viewpoints that surround the object in a dome.

We created a dataset of images and viewpoints resembling the open source GQN dataset provided in [4]. In the Gazebo simulator, an object is spawned in the middle of a circumference with a 0.2-m radius, and is positioned on top of a table and closer to the left arm of the Baxter robot, rather than the middle, to enable the left arm to surround the object as much as possible. We selected 101 objects from the 3DGEMS dataset that were individually spawned on top of the table in front of the Baxter robot. Images were captured for 20 viewpoints that sitting along the 0.2 m radius circumference, for each object spawned on the table. Image and viewpoint data were paired together in the same format as the GQN dataset.

5 Results

Essentially, the GQN predicts an output image from a given query viewpoint; the uncertainty of these predictions can be observed in the predicted images as blurred pixels. As is mentioned in Sect. 3, we quantify this uncertainty by calculating the pixel-wise variance of multiple sampled predictions. Areas in the learning environment that show high uncertainty are defined as pixels with high values of variance.

An overview of the network model is shown in Fig. 1, in which an observed image is paired with V_i^1 which represents the observed viewpoint. This information is then encoded into a neural scene representation r. V_i^3 represents a query viewpoint(unobserved) for the model to predict. The uncertainty map generated from the pixel-wise variance can be displayed as grayscale images to observe areas of high uncertainty. Whiter pixels represent higher values of pixel variance.

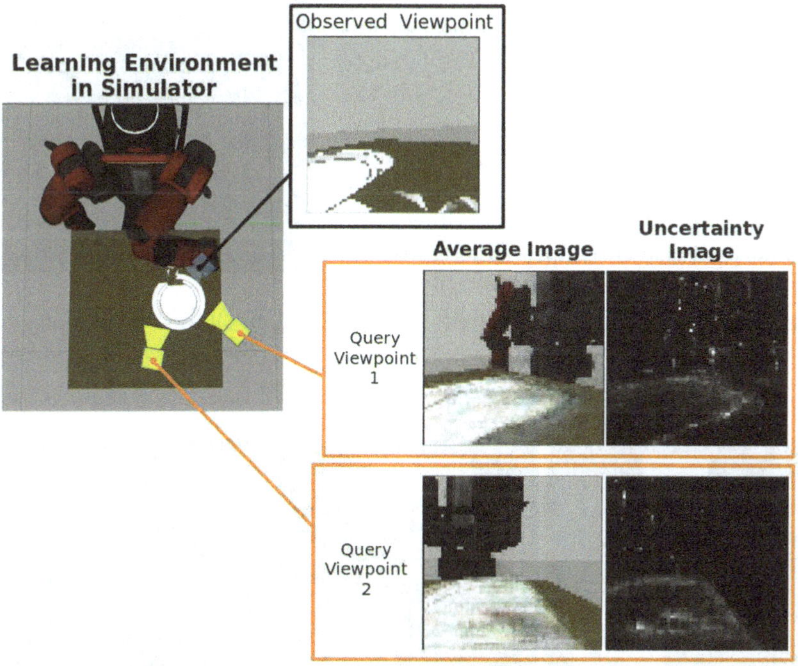

Fig. 5. Average of 100 predictions and uncertainty image created from predictions (within the orange bounding box) of two query viewpoints, that are generated from one observed viewpoint.

In Fig. 5, the learning environment shown on the left consists of the Baxter robot, a table, and a broken plate positioned on top of it. The blue camera icon represents the direction of the observed viewpoint, and the yellow camera icons represent the direction of query viewpoints. A scene representation is created from the observed viewpoint and image, and from that representation vector, we generate 100 predictions and render their average, as well as the uncertainty map calculated by the pixel-wise variance of the predictions. An uncertainty map of the environment is the accumulation of uncertainty images. The images from the uncertainty map shown in Fig. 5 are grayscale images of the pixel-wise variance calculated from the rendered predictions. Whiter pixels are more concentrated toward the object on the table because the learning environment always consists of the Baxter robot and table in constant positions whereas the object on top of the table changes. Increasing the number of observed images will increase the certainty of predictions and therefore reduce the number of whiter pixels observed on the uncertainty map.

6 Discussion

The angle which observes highest variance should logically be the observed viewpoint. However, this is not the case in the example shown in Fig. 3. As shown in Fig. 6, there are 3 different Shepard-Metzler objects in the observations. At this particular query viewpoint, the model has difficulty in generating an accurate prediction as shown under the neural rendering of each observed object due to similarity between shapes. Therefore, the observed viewpoint will not necessarily be the viewpoint with the lowest uncertainty, and the model will require observation of more images to generate an accurate prediction.

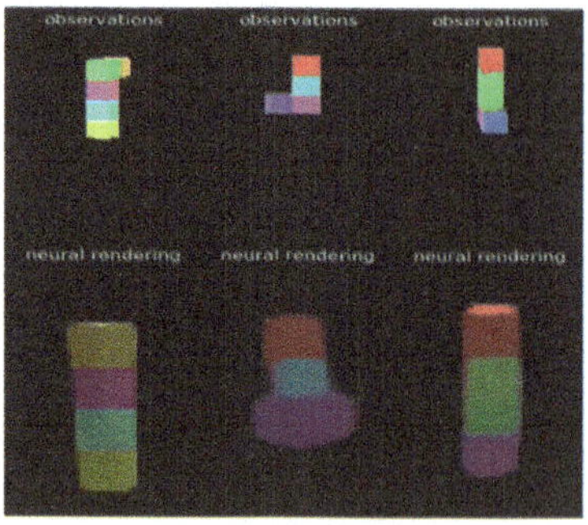

Fig. 6. Objects with similar structure causes ambiguity in predictions.

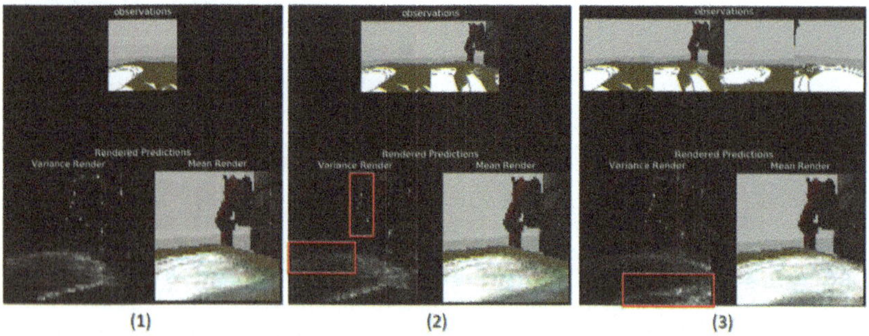

Fig. 7. Increase in the number of observations observed by the model decreases the number of white pixels in the uncertainty image. (1) shows an uncertainty image generated from one observation. In (2) the model has 2 observations, and areas within the red bounding box show a decrease in white pixels as compared to (1). Providing 4 observations as shown in (3) may change the area of concentrated white pixels as shown within the red bounding box.

As described in Sect. 5, increasing the number of observed images will reduce the number of whiter pixels that are observed on the uncertainty map due to increase in certainty. A qualitative observation of the generated uncertainty images shown in Fig. 7 demonstrate the decrease in uncertainty with increase in number of observations. We described the relation between the model's uncertainty and the number of white pixels shown in generated uncertainty images. In Fig. 7, the image in (2) demonstrates the model's decrease in uncertainty from (1) as shown by the decrease in white pixels in areas enclosed by the red bounding boxes, and this is due to the additional observation given to the model. The image in (3) shown in Fig. 7, shows the uncertainty image generated when 4 observations are obtained. The difference in (2) and (3) show that the model had a misconception about the white object on the table. If we observe the prediction in (2), the model predicts the object has an edge right in front of the camera, but the additional 2 observations shows the model that the object is bigger than it originally predicted, and therefore we observe the shift in uncertainty towards the bottom edge of the image. As shown within the red bounding box in (3), the white pixels no longer form a straight line where the model predicted and edge in (2), and white pixels become more concentrated towards the bottom of the image.

7 Conclusion

In this study, we demonstrate a method for viewpoint planning by creating an uncertainty map of the learning environment. A GQN creates a scene representation vector from the image input, from which we sample 100 predictions and then calculate the pixel-wise variance of predictions to generate an uncertainty map of the learning environment. We define an area of high uncertainty in the learning environment as pixels that have a high variance value. The viewpoint planner determines the next suitable viewpoint by means of a hill–climbing method to find a viewpoint that captures an image that features the highest overall variance represented in the predictions.

7.1 Future Work

We intend to apply the uncertainty map discussed herein to the simultaneous operation of two robot arms. A single arm will move an attached camera to search for an optimal viewpoint, and the other arm will carry out the designated task based on information collected from the observed viewpoint.

Acknowledgment. This work was supported by Japan Science and Technology Agency (JST) CREST Grant Number JPMJCR15E3 "Symbol Emergence in Robotics for Future Human-Machine Collaboration", Japan.

References

1. Bajcsy, R.: Active perception. Proc. IEEE **76**(8), 966–1005 (1988)
2. Bohg, J., Morales, A., Asfour, T., Kragic, D.: Data-driven grasp synthesis-a survey. IEEE Trans. Rob. **30**(2), 289–309 (2014)
3. Chen, S., Li, Y., Kwok, N.: Active vision in robotic systems: a survey of recent developments. Int. J. Robot. Res. **30**, 1343–1377 (2011)
4. Eslami, S.M.A., et al.: Neural scene representation and rendering. Science **360**(6394), 1204–1210 (2018)
5. Gregor, K., Danihelka, I., Graves, A., Wierstra, D.: DRAW: a recurrent neural network for image generation. CoRR, abs/1502.04623 (2015)
6. Kappler, D., Bohg, J., Schaal, S.: Leveraging big data for grasp planning. In: 2015 IEEE International Conference on Robotics and Automation (ICRA), pp. 4304–4311 (2015)
7. Lenz, I., Lee, H., Saxena, A.: Deep learning for detecting robotic grasps. Int. J. Robot. Res. **34**(4–5), 705–724 (2015)
8. Levine, S., Pastor, P., Krizhevsky, A., Ibarz, J., Quillen, D.: Learning hand-eye coordination for robotic grasping with deep learning and large-scale data collection. Int. J. Robot. Res. **37**(4–5), 421–436 (2018)
9. Mahler, J., et al.: Dex-Net 2.0: deep learning to plan robust grasps with synthetic point clouds and analytic grasp metrics (2017)
10. Mahler, J., Matl, M., Liu, X., Li, A., Gealy, D., Goldberg, K.: Dex-Net 3.0: computing robust robot suction grasp targets in point clouds using a new analytic model and deep learning. arXiv preprint arXiv:1709.06670 (2017)
11. Mahler, J., et al.: Learning ambidextrous robot grasping policies. Sci. Robot. **4**(26), eaau4984 (2019)
12. Manuelli, L., Gao, W., Florence, P., Tedrake, R.: kPAM: keypoint affordances for category-level robotic manipulation. arXiv, abs/1903.06684 (2019
13. Nguyen-Phuoc, T., Li, C., Theis, L., Richardt, C., Yang, Y.: HoloGAN: unsupervised learning of 3D representations from natural images. CoRR, abs/1904.01326 (2019)
14. Qiu, Y., Satoh, Y., Suzuki, R., Iwata, K., Kataoka, H.: Multi-view visual question answering with active viewpoint selection. Sensors **20**(8), 2281 (2020)
15. Rasouli, A., Tsotsos, J.K.: The effect of color space selection on detectability and discriminability of colored objects. CoRR, abs/1702.05421 (2017)
16. Sather, J., Zhang, X.J.: Viewpoint optimization for autonomous strawberry harvesting with deep reinforcement learning. arXiv, abs/1903.02074 (2019)
17. Sitzmann, V., Zollhöfer, M., Wetzstein, G.: Scene representation networks: continuous 3D-structure-aware neural scene representations. CoRR, abs/1906.01618 (2019)
18. Tobin, J., OpenAI, Abbeel, P.: Geometry-aware neural rendering. arXiv, abs/1911.04554 (2019)

Active Learning-Based Data Collection in Crowd Replication

Lulu Gao[1](✉) and Shin'ichi Konomi[2]

[1] Graduate School of Information Science and Electrical Engineering, Kyushu University, Fukuoka, Japan

[2] Faculty of Arts and Science, Kyushu University, Fukuoka, Japan

Abstract. Crowd replication, which combines crowd sensing, direct observation, and mathematical modeling to enable efficient and accurate evaluation of crowd, is a low-effort, easy-to-adopt, and cost-effective mechanism for crowd data collection and analysis. In crowd replication, the quality of data collection is particularly important, which depends on the representativeness of the target population-based sampling. The main two target selection strategies, population-based sampling strategy, and cluster sampling strategy will be labor-intensive and time-consuming to obtain the stable, reliable, and valid data. Therefore, in this paper, a novel method of data collection in crowd replication based on active learning, which is a modern method in machine learning, aiming to reduce the sample size, complexity, and increase the accuracy of the data tasks as much as possible with less data is proposed. We apply active learning to allow us to obtain the dataset with high representativeness and informativeness. We demonstrate with experimental results that, compared with the traditional probability-based sampling strategies, the more representative samples and dataset can be stably captured by our contribution.

Keywords: Crowd replication · Data collection · Active learning

1 Introduction

With the development of wireless communication technology and sensor technology and the widespread popularity of wireless mobile terminal devices, mobile devices, which integrate more and more sensors and have more and more powerful abilities of computing and sensing [1–3]. With the large sensing capabilities of mobile devices, many people can use their mobile devices to sense different aspects of themselves and the environment, which can enable a broad range of applications, such as a smart city, healthcare, mobile social networks, and advertising if people's sensing data are reliable and valid. However, due to the resource limitations, including the constraints of mobile devices and resource cost, and the issue of privacy, security and data integrity, collecting reliable and valid data is very difficult [4, 5].

Crowd replication was proposed as a sensor-assisted approach for crowd data collection and analysis, which can easily get accurate fine-grained data in terms of human

K. Yada et al. (Eds.): JSAI 2020, AISC 1357, pp. 49–60, 2021.
https://doi.org/10.1007/978-3-030-73113-7_5

activities and movement trajectories and discover subtle patterns of human behaviors [5]. In crowd replication, the researcher carrying some mobile devices when replicating the target, they can simultaneously effectively capture the reliable and valid data of target corresponding to the observed behaviors [6]. There are three main advantages in crowd replication: better control the sample selection to collect reliable data, use off-the-shelf mobile devices (already install some sensors or some applications) to significantly reduce resource limitations and protect the security and privacy of people.

The sufficient accuracy and the representativeness of the motions of people of the collected data is the essential foundation for crowd replication. To ensure that, the target is randomly selected based on population in crowd replication [6]. Although it can be more accurate for single data due to the precise hardware or software, it does not necessarily reflect the true data distribution, which means unreliable and invalid for the entire dataset. As a method of constructing an effective dataset in machine learning, active learning can find informative samples for the whole dataset according to the historical data, thereby improving the reliability and accuracy of the dataset within the limited resources [7]. Active learning employs candidate selection strategies that query the informativeness of the samples for the current task via uncertainty measurement, thus the data of a certain sample that is most beneficial to the task is collected. Therefore, we proposed a novel method of data collection based on active learning in which each time a target is selected, which makes the whole dataset more reliable.

The proposed method is an efficient method for data collection of crowd replication which is a low-effort, easy-to-implement, and cost-effective measure to quantify human motions. In summary, the contributions of this paper are as follows: 1) The current methods used in the data collection of crowd replication are analyzed, and the problems they face are pointed out. 2) A data collection method based on active learning applied to crowd replication is proposed. 3) In this paper, the population characteristics of the data and the informativeness of the sample are considered at the same time when selecting the target. 4) The performance of the method proposed in this paper and the existing methods are compared to verify the efficiency of the former one from comprehensive experiments.

2 Related Works

2.1 Crowd Replication

Crowd replication was proposed by Hemminki et al. as a sensor-assisted approach for quantitative evaluation, which was used to collect the dataset through surveys and direct observation, of the sociability, activities and human behavior in public spaces [5]. Time-consuming and laborious are the main limitation of the traditional methods of surveys and direct observation to obtain data, which often leads to a small sample size and insufficient dataset. Crowd replication combines mobile sensing, direct observation to collect a large fine-grained dataset of human behaviors with significantly smaller resources than visual tracking or other pervasive sensing solutions. Crowd replication could migrate the privacy and security issues of mobile sensing by replicating the movements of the target sample within target space by researchers and enable efficient and accurate quantitative evaluation of public spaces and humans by mathematical modeling. The main idea in crowd replication is to enhance direct observation, in which researchers can carry some mobile devices equipped with sensors and applications to simultaneously effectively capture the detailed motions information and spatiotemporal information of target corresponding to the observed behavior, so that more fine-grained data can be collected.

The entire crowd replication process can be divided into three parts: study preparation, data collection, and data analysis [6]. Study preparation is to do a good survey to determine the target public space and the method of sampling which should capture the representative subset of the target population and representative behaviors via population-based sampling strategy and cluster sampling strategy. Data collection is the implementation of the previously developed plan by researchers, replicating people' s motions on a general level and collecting fine-grained data and avoiding interfering with normal behavior within the space. Finally, the collected dataset will be processed in data analysis, and the target problem will be analyzed in combination with the processed data to give a reasonable explanation.

2.2 Active Learning

Active learning is a modern method in machine learning, aiming to reduce the sample size, complexity, and increase the accuracy of the data tasks as much as possible with fewer data. The key hypothesis of active learning is that the learning mechanisms will be more intelligent if the learning algorithm can actively choose the most significant unlabeled data. An active learner will query only a small number of valuable unlabeled instances to be labeled by an oracle or annotator to automatically enlarge the labeled dataset, in an intelligent manner [7, 20].

There are three main scenarios that have been studied of active learning, *member-ship query synthesis*, *stream-based selective sampling*, and *pool-based sampling* [13, 17, 18]. In query synthesis, any unlabeled instance can be queried by an active learner, including the model-generated although it may have no practical meaning and cannot be labeled by a human annotator. The unlabeled instances will be query sequentially by the learner in stream-based selective sampling [14]. And the learner will decide whether the instance be annotated or not. In pool-based sampling, many unlabeled instances are assumed to be available. In this kind of scenario, the learner should rank the entire unlabeled instances according to an informativeness measure, that is the pool of unlabeled instances, and then, query the most informative one [14]. The main difference between pool-based sampling and stream-based selective sampling is that the former should evaluate all unlabeled data before select query, while the latter just query the instance in sequence [16].

The measure of informativeness evaluation is vital in all active learning scenarios and can be parted into *uncertainty selection, query by committee, expected objective change,* and *data-centered method* [15, 16, 20]. Uncertainty selection would select the most uncertain instance on the prediction of the current model to query. Query by committee will query the most disagreeing instance of the committees' prediction. Each committee member is a different model based on the current training set. Expected objective change, which query the instance that could make the maximizing impact on the objective. For example, maximizing model change, maximizing the generalization error reduction, maximizing the output variance reduction. The most representative of the most informative instance will be queried in the data-centered method.

2.3 Population-Based and Cluster Sampling

There is less research on crowd replication, which mainly used for data collection and offline data analysis of the crowd. Population-based sampling selection and cluster sampling are the main two selection strategies in crowd replication [6]. Population-based sampling strategy collects data based on the relevant population characteristics within the target area, therefore, the representativeness of the collected data could be ensured. Cluster sampling is interpreted as the physical entry points of the space as clusters, in which random sampling is applied, which could simultaneously get the representativeness of both target entry and target user's behavior when a large amount of data is obtained [8].

The existing selection strategy will be close to the population distribution within a certain target area in the case of a large amount of collected data, but a large amount of data means time-consuming and labor-intensive. What's more, a random selection is made each time, the final population distribution of all the samples is unstable, meaning that it may take multiple times to sample in the target area to obtain the stability of the data and the stability of the analysis results.

3 Active Learning-Based Crowd Data Collection

The intuition of our proposal is that collecting data closer to the population distribution each time allows us to obtain a more reliable and valid dataset.

3.1 Jensen-Shannon Divergence

Many dissimilarity functions based on information theory that can measure the distance of two probability distributions, that is the similarity, are extensively studied [9]. Among the proposed functions, Jensen–Shannon divergences (JS divergence) is one of the best known. JS divergence is based on the Kullback–Leibler divergence (KL divergence), with some useful improvement, including that it is symmetric between two distributions and it always has a finite value [19]. For discrete probability distributions, P and Q defined on the same probability space, χ, the KL divergence of Q from P is defined to be:

$$D_{KL}(P\|Q) = -\sum_{x\in\chi} P(x)\log\frac{P(x)}{Q(x)} = -\sum_{x\in\chi} P(x)\log\frac{Q(x)}{P(x)} \tag{1}$$

and the JS divergence is defined to be [10]:

$$JS(P\|Q) = \frac{1}{2}KL(P\|M) + \frac{1}{2}KL(Q\|M), \text{ where } \mathrm{M} = (\mathrm{P}+\mathrm{Q})/2. \tag{2}$$

The JS divergence can be generalized to provide such a measure for any finite number of distributions [9]. This is useful for the study of decision problems with more than two classes involved. Higher JS divergence value means a larger dissimilarity of two distributions. On the contrary, a smaller JS divergence indicates closer distributions, and zero represent two identical distributions. Thus, it has been applied in bioinformatics and genome comparison, protein surface comparison, the social sciences, the quantitative study of history, machine learning and active learning to measure the utility of sample [11, 12, 21]. Therefore, we can apply it in crowd replication to reduce the distribution divergence between the true dataset and the collected dataset during the process of data collection and get reliable and valid results.

3.2 Active Learning-Based Crowd Data Collection

We could consider the process of data sampling as the process of changing the distribution of the collected dataset because when we select the target to collect behavior data and add the collected data to the collected dataset, the distribution of the collected dataset will be changed. According to certain features, P_c is the distribution of the current collected sample dataset, and P_r is the actual distribution of the population. To ensure that the sample dataset we collected could be the representative subset of the population, and the basis of the analyzed result is reliable and valid, we should try to minimize the divergence between P_c and P_r, $Div(P_c, P_r)$. Let D be the currently collected dataset and $\mathbb{x}_i$be the sample with the value x_i which is one of the value of discrete probability variable χ, that is $\{x_1, \ldots, x_n\}$, according to the same certain features.

Now, our intent is to find x^* that could minimize $D(P_c, P_r)$, that is:

$$\min_{x_i} ||Div(P_c^i(D \cup \mathbb{x}_i), P_r)||, x_i \in \chi \tag{3}$$

where $P_c^i(D \cup \mathbb{x}_i)$means the new P_c after adding the sample $\mathbb{x}_i$with the values of x_i to D. The solution to this optimization problem is iteratively select the value x_i to compare the difference of new P_c^i and P_r, and then collect the sample $\mathbb{x}^*$with the value x^* to obtain the reliable and valid dataset. The procedures are outlined in Table 1.

Table 1. Active learning-based data collection in crowd replication

Input: actual distribution P_r, the values of χ: $\{x_1, x_2, \ldots, x_i, \ldots, x_n\}$
Output: the value list we should collect
repeat
for each $x_i \in \{x_1, x_2, \ldots, x_i, \ldots, x_n\}$ do
Computing and saving the $Div(P_c^i(D \cup \mathbb{x}_i), P_r)$
end for
solving the optimization problem $\min_{x_i} ||Div(P_c^i(D \cup \mathbb{x}_i), P_r)||$
Appending the optimal solution x^*
until resource limitations

Active learning can actively select the next sample, when it is used in data collection, it can make the distribution of the collected dataset closer to the population distribution, obtaining more reliable dataset. As can be seen from the algorithm, each cycle, it will iterate all values of χ, to select the value that can achieve the smallest divergence between P_c and P_r, which is also the informative sample value for current situation. According to the output of the algorithm, the value list, to select the target and obtain the motion data by replication, the representative samples can be ensured to a large extent. This is exactly the stream-based sampling strategy in active learning and can let us get a dataset with high informativeness, reliability and validity.

For example, we want to quantitatively evaluate the specific public space and to implement the goal, we could know about people's activities in the target area. In this scenario, the sampling distribution close to the population distribution could ensure that the dataset we collected is reliable and valid to a certain extent. There are many characteristics to distinguish the crowd (gender, age, etc.), while we concentrate on the general feature of gender. Then χ is the gender of the sampling target each time, and the values can only be female and male. P_r is the real gender ratio in the target public area we investigated via the historical data. P_c is the current gender distribution of the sampled dataset D. When it is changed, it is necessary to determine which value of gender (male, female) of the next sample added to D could make the gender ratio P_c of the updated collection dataset $D \cup \mathbb{x}$is closer to P_r, which is to solve the optimization problem of $\min_{x_i} ||Div(P_c^i(D \cup \mathbb{x}_i), P_r)||$. . As the sampling process carrying on, our costs will continue to increase until reaching our expected costs. In this way, the gender of each target we selected to replicate the movement and collect the fine-grained data allow us to get a dataset whose gender distribution is closer to the population, which could be said that the dataset we collected is representative and the results of our quantitative evaluation of public space are reliable and valid.

4 Experiment

4.1 Scenario

Population-based sampling strategy and cluster sampling strategy of target selection in data collection of crowd replication are all used in the scenario showed in [6]. Therefore, for the convenience of our comparison, we choose the same scenario that is to replicate the selected representative target according to different strategies and collect crowd activities from the devices equipped with different sensors and user-friendly application carried by researchers to quantitatively evaluate public space to conduct our experiments. While due to resource constraints, we can just simulate the scenario on a computer (Intel i7-5557U 3.1 GHz, 16G, 500 GB). The detailed experiment can be described as, in a public space with four entrances and exits (such as a park with multiple entrances and exits), a researcher collects corresponding data via carefully replicate and record representative crowd behaviors of different entrances. In the experiment, we need to select the target person to record the activities, which uses cluster sampling strategy in [6] (now, we just consider the gender feature to divide the population) at one entrance selected by population-based sampling strategy in [6] from the four entrances and exits to collect data. And the JS divergence of the gender distribution between the collected dataset and the population could be the measure of the reliability and validity of the collected dataset.

There are two tasks in our current experimental scenario. One is to select the entrance and exit and then to choose a suitable target person from the current openings to replicate to collect the behavior data. In the first task, the selection of the entry and exit, which adopts the population-based strategy traditionally, needs to meet the distribution of the total number of people at four entrances and exits to obtain a representative crowd. After that, in the second task, the target sample that meets the gender ratio of the population of the currently selected entrance and exit is picked up for replicating and collecting information to obtain informative sample data, which employ the cluster sampling strategy previously.

We could compare the performance of the proposed method and the existing method via the observation of the changes in the distribution of gender selection for each entry and each target person.

4.2 Validity

Validity can be understood as the representativeness of the dataset. For the experimental scenario, the validity of the collected dataset of the selected target can be regarded as the representativeness of the population, which is the ability to represent the population. From the perspective of gender alone, it is whether the gender distribution of the obtained data can represent the population distribution. It can be explained by observing the JS between the distribution of the collected data and the population distribution. The smaller the difference is, the closer the two distributions are, the greater the capacity of the collected data to represent the population. For example, if we want to investigate people's view on the price of a particular vegetable in the vegetable market. To make our data more representative and obtain more persuasive analysis, we should investigate more women than men (there will be more women in the vegetable market only consider the gender of people in the vegetable market).

Figure 1 shows the changes in the JS divergence of the distribution between the sampled data and the population in 100 sample selection. Figure 1(a) is the JS divergence changes of entry selection, and Fig. 1(b) is the JS divergence changes of target gender distributions in each entry with the experiment conducted on. From Fig. 1(a), we can see that compared with population-based sampling strategy, proposed method can obtain more accurate data, that is, the ratio of each entrance selection is close to the ratio of the total number of people in the four entries. The same situation also occurs in the gender ratio change of the sample for each entry in Fig. 1(b), which we can get representative samples and informative data.

We use the same source data and different source data respectively to do 100-time sampling, each time collecting 100 samples (including entry selection and target sampling) and observe the changes in JS differences of different measures. Figure 2 and Fig. 3 show that no matter the same source data and different source data are, our method can achieve better performance than the traditional, which could obtain the more representative and more valid dataset closer to the population and closer to the fact, leading to accurate analysis results.

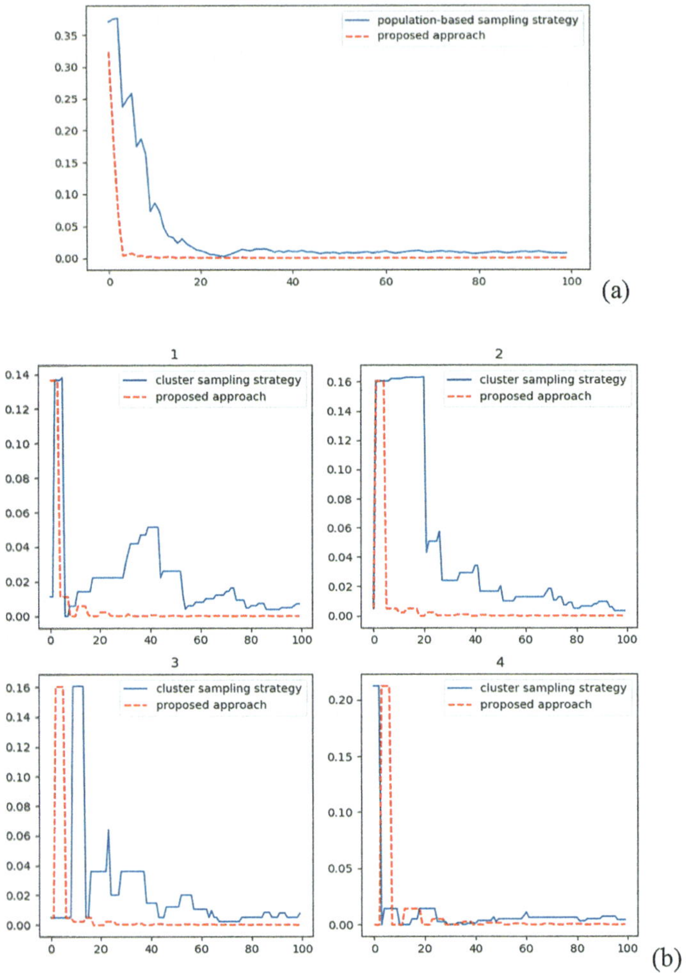

Fig. 1. Changes in the JS divergence of the distributions between the samples and the population in 100 sample selection.

4.3 Reliability

To some extent, the validity of the dataset, which is explicitly explained above, is also part of the reliability. Here the reliability is illustrated just from two aspects: believability and stability. Believability means whether the collected data is credible and whether credible data can be obtained under different circumstances. And what the stability mainly considered is no matter what the situation, whether the valid and credible dataset can be collected stably.

Figure 1 shows that although the two strategies and the proposed method can get a dataset close to the population distribution as the experiment continues, the proposed method can collect a more valid and effective dataset faster. That is, the proposed can ensure an accurate dataset to be obtained for a longer time. Even if we have to terminate

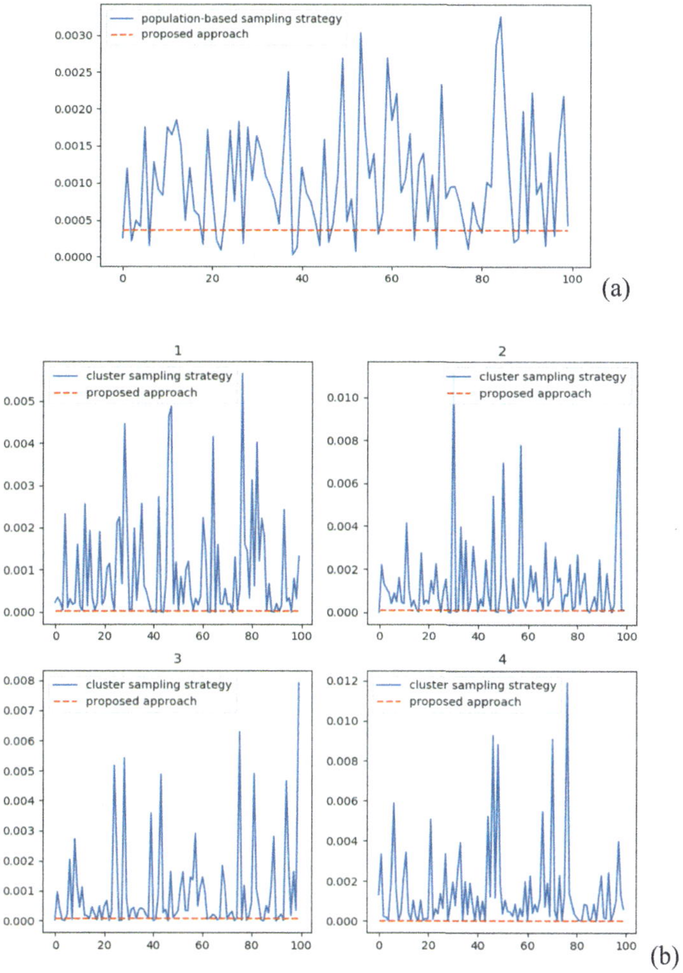

Fig. 2. Using the same source data to collect the dataset 100 times, getting the sample distribution each time and computing JS divergences between the population distribution

the experiment in the face of emergencies, we can ensure that the sample data we already obtained is reliable. It also reflects the credibility of the data sampling because the termination of the experiment generates different situations, while the proposed measure still able to obtain relatively accurate dataset.

Figure 2 depicting the difference in JS values sampled 100 times using the same source data, and Fig. 3 describes the difference in JS values tested 100 times using different source data. Both show the reliability of the proposed method. It can be seen from Fig. 2 that every time the same source data is encountered, the proposed method can collect a relatively good data set. Sampling strategies based on population sampling and cluster sampling will unexpectedly collect better data sets with irregularities. Therefore, the proposed method is stable, that is, more reliable. Figure 3 shows that for different

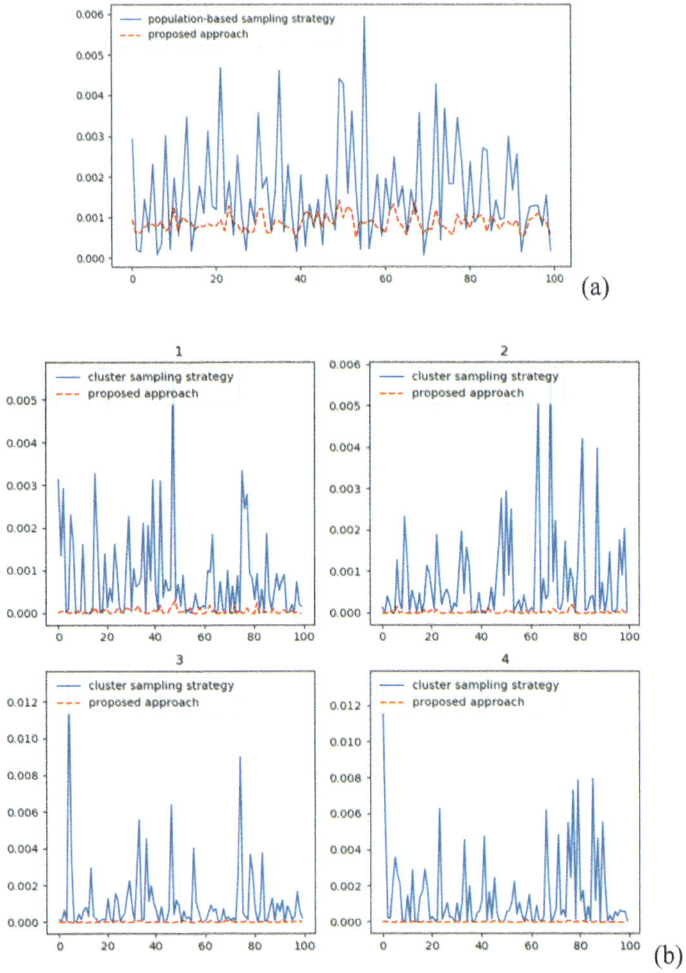

Fig. 3. Using the different source data to collect the dataset 100 times, getting the sample distribution each time and computing JS divergences between the population distribution

source data, using the suggested method, we can obtain a better collection of datasets. In Fig. 3(a), using the proposed method, the JS divergence of the sampling rate of different items is lower than the JS divergence of using the population-based sampling strategy, indicating that we can get appropriate entry closer to the population. Similarly, in Fig. 3(b), the gender ratio of the specific items we collected is more relative to the population distribution. We can see that for different source data, the proposed method can always obtain more valid data sets than traditional methods, which means more representative datasets can be credibly collected.

5 Discussion and Future Work

In this work, we proposed a new crowd data collection in crowd replication, active learning-based data collection in crowd replication, that employs the active learning technology to actively select the sample to complicate and collect fine-grained human behavior data. The experiment results show that the performance of the proposed method is better than the existing strategies in terms of validity and reliability that stably ensure a valid dataset can be obtained. Because every single data collected of the dataset in this method will get to make the distribution of the collected dataset close to the population's, leading to an effective representative dataset.

Our future work includes some improvements to the approach. There are many features to divide the population, and it is not sufficient to select the simplest gender feature to classify the people in the experiment, more features to distinguish the crowd will be studied. The experiment scenario in this paper is too simple because the number of people in different regions may have different changes, such as subway stations and bus stops, hence more environmental factors should be considered. Besides, the evaluation of historical data is not necessarily precise, and the distribution close to historical data may not be perfect. A careful survey should be carried on problems to get more useful information and more accurate statistics.

Acknowledgement. This work was supported by JSPS KAKENHI Grant No. JP20H00622, JP17KT0154.

References

1. Hong, H., Chaocan, X., et al.: Survey on crowd-sensing networks. J. Jilin Univ. (Inf. Sci. Ed.) **34**(3), 374–383 (2016)
2. Gaire, R., Ghosh, R.K., Kim, J., Krumpholz, A., Nepal, S.: Crowdsensing and Privacy in smart city applications. In: Smart Cities Cybersecurity and Privacy. Elsevier (2019)
3. Alkhelaiwi, A.T.: Crowd-sensing for smart city applications: towards solving crowd-sensing data challenges by introducing edge and cloud services. Ph.D. thesis, University College Cork (2019)
4. Ganti, R.K., Ye, F., Lei, H.: Mobile crowdsensing: current state and future challenges. IEEE Commun. Mag. **49**(11), 32–39 (2011)
5. Hemminki, S., Kuribayashi, K., Konomi, S., Nurmi, P., Tarkoma, S.: Quantitative evaluation of public spaces using crowd replication. In: Proceedings of the 24th ACM SIGSPATIAL International Conference on Advances in Geographic Information Systems (SIGSPACIAL 2016), Article 63, pp. 1–4. ACM, California (2016)
6. Hemminki, S., Kuribayashi, K., Konomi, S., Nurmi, P., Tarkoma, S.: Crowd replication: sensing-assisted quantification of human behavior in public spaces. ACM Trans. Spat. Algorithms Syst. **5**(3), 1–34 (2019)
7. Settles, B.: Active learning literature survey (2010)
8. Wang, J.-F., Stein, A., Gao, B.-B., Ge, Y.: A review of spatial sampling. Spat. Stat. **2**, 1–14 (2012)
9. Lin, J.: Divergence measures based on the Shannon entropy. IEEE Trans. Inf. Theory **37**(1), 145–151 (1991)

10. Fuglede, B., Topsoe, F.: Jensen-Shannon divergence and Hilbert space embedding. In: Proceedings of the International Symposium on Information Theory, Chicago, IL, USA. IEEE (2004)
11. Simon, D.D., Robert, H., Sara, K., Tim, H.: Bootstrap methods for the empirical study of decision-making and information flows in social systems. Entropy **15**(6), 2246–2276 (2013)
12. Goodfellow, I.J., et al.: Generative adversarial nets. In: Proceedings of the 27th International Conference on Neural Information Processing Systems, (NIPS 2014), vol. 2, 2672–2680. MIT Press, Cambridge (2014)
13. Li, K., Zhou, X., Lin, F., Zeng, W., Alterovitz, G.: Deep probabilistic matrix factorization framework for online collaborative filtering. IEEE Access **7**, 56117–56128 (2019)
14. Liu, D., Liu, Y.: An active learning algorithm for multi-class classification. Pattern Anal. Appl. **22**(3), 1051–1063 (2019). https://doi.org/10.1007/s10044-018-0716-1
15. Bernard, J., Zeppelzauer, M., Lehmann, M., Müller, M., Sedlmair, M.: Towards user-centered active learning algorithms. Comput. Graph. Forum **37**(3), 121–132 (2018)
16. Hao, S., Peiying, H., Zhao, P., Hoi, S., Miao, C.: Online active learning with expert advice. ACM Trans. Knowl. Discov. Data **12**(5), 1–22 (2018). https://doi.org/10.1145/3201604
17. Xu, Q., Zheng, R.: When data acquisition meets data analytics: a distributed active learning framework for optimal budgeted mobile crowdsensing. In: Proceedings of IEEE Conference on Computer Communications (IEEE INFOCOM 2017), Atlanta, GA, pp. 1–9 (2017). https://doi.org/10.1109/INFOCOM.2017.8057034
18. Settles, B.: Active learning literature survey. University of Wisconsin-Madison Department of Computer Sciences (2009)
19. Wang, Y., et al.: Spectral clustering based on JS-divergence for uncertain data. In: 2017 IEEE International Conference on Systems, Man, and Cybernetics (SMC), Banff, AB, pp. 1972–1975 (2017). https://doi.org/10.1109/SMC.2017.8122907
20. Konomi, S., Gao, L., Mushi, D.: An intelligent platform for offline learners based on model-driven crowdsensing over intermittent networks. In: Rau, P.-L.P. (ed.) HCII 2020. LNCS, vol. 12193, pp. 300–314. Springer, Cham (2020). https://doi.org/10.1007/978-3-030-49913-6_26
21. Melville, P., Yang, S., Saar-Tsechansky, M., Mooney, R.: Active learning for probability estimation using Jensen-Shannon divergence. In: Gama, J., Camacho, R., Brazdil, P.B., Jorge, A.M., Torgo, L. (eds.) ECML 2005. LNCS (LNAI), vol. 3720, pp. 268–279. Springer, Heidelberg (2005). https://doi.org/10.1007/11564096_28

Mining in Discharge Summaries

Shusaku Tsumoto[1(✉)], Tomohiro Kimura[2], and Shoji Hirano[1]

[1] Department of Medical Informatcs, Faculty of Medicine, Shimane University, 89-1 Enya-cho, Izumo 693-8501, Japan
{tsumoto,hirano}@med.shimane-u.ac.jp

[2] Medical Services Division, Faculty of Medicine, Shimane University, 89-1 Enya-cho, Izumo 693-8501, Japan
t-kimura@med.shimane-u.ac.jp

Abstract. This paper proposes a method of construction of discharge summaries classifier. First, morphological and correspondence analysis generates a term matrix from text data. Then, machine learning methods are applied to a term matrix. The method compared several machine learning methods by using discharge summaries stored in hospital information system. The experimental results show that random forest is the best clasifier, compared with deep learning, SVM and decision tree.

1 Introduction

Computerization of patient records enables to store "big unstructured text data" in a Hospital Information System (HIS). For example, our system in Shimane university hospital, where about 1000 patients visit outpatient and about 600 patients stays inpatient, additionally stores about 200GB text data per year, including patient records, discharge summaries and reports of radiology and pathology. Application of text mining to these resources is very important to discover useful knowledge from text data, which can be viewed as a new type of support of clinical actions, researches and hospital management.

This paper proposes a method for construction of classifiers for discharge summaries, which consists of the following four steps. First, morphological analysis is applied to a set of summaries and a term matrix is generated. Second, correspond analysis is applied to the term matrix with the class labels, and two dimensional coordinates are assigned to each keyword. By measuring the distance between categories and the assigned points, ranking of key words for each category will be generated. Then, keywords are selected as attributes according to the rank, and training example for classifiers will be generated. Finally learning methods are applied to the training examples. We conduct experimental validation with four methods, random forest, deep learning (multi-layer perceptron), SVM and decision tree induction method. The results show that random

This research is supported by Grant-in-Aid for Scientific Research (B) 18H03289 from Japan Society for the Promotion of Science(JSPS).

K. Yada et al. (Eds.): JSAI 2020, AISC 1357, pp. 61–72, 2021.
https://doi.org/10.1007/978-3-030-73113-7_6

forest achieved the best performance and the second best was the deep learner with a small difference, but decision tree methods with many keywords performed only a little worse than neural network or deep learning methods. The paper is organized as follows. Section 2 explains our motivation. Section 3 gives a proposed mining process. Section 4 shows the experimental results. Section 5 discusses the results obtained. Finally, Sect. 6 concludes this paper.

2 Motivation

The principal purpose of AI application to hospital data is to make the work of medical staff more efficient in clinical environment. One of the laborious work for doctors and nurses is documentation, including detailed description of records, which is needed for several purposes, such as medical payment submission, information exchange between hospitals or clinics. The validity of many documents should be checked, mainly because most of the medical payment are based on submission of medical fee statement, whose information is extracted from medical documents. Especially, in large-scale hospital in Japan, we need to submit statements following the system of Diagnosis Procedure Combination (DPC) [1]. DPC is assigned to a disease for which the major parts of medical resources were used during hospitalization of a patient. For each day of hospital stay, a payment point is assigned for each DPC code. Thus, medical payment by DPC depends on the length of hospital stay and diagnosis and medical procedures, and differs from the traditional payment system, which depends on a piece rate of each medical action.[1] Since DPC codes are used to describe each hospitalization in hospital information system from the viewpoint of medical payment, the assigned code may differ from diseases of medical interests, which makes doctors difficult to assign DPC code.

Thus, medical information managers have to check whether DPC code is correct or not, according to the clinical records before submission of statements. Mainly, they check the validity of assigned DPC codes by using discharge summaries and patient records. For example, in Shimane University Hospital, 40 patients are discharged every day in average. In one month, about 1200 patients are discharged, which means that medical information managers have to check 1200 discharge summaries and patient records for one month, where efficient checking is very important. In this hospital, six managers are working for checking patient records and DPC codes.

Thus, automated document classification with correct DPC code will help medical information managers for submission of accurate fee statements in large-scale hospital, which enables them to focus on checking complicated cases

[1] Outpatient clinic is still based on the action-based payment system even in large hospitals.

3 Methods

3.1 Discharge Summary

A discharge summary includes all clinical processes during the hospitalization, whose written style is formal, compared with regular records. Thus, conventional text mining approach can be used to extract enough keywords from each text.

3.2 Motivation for Feature Selection

It is well known that feature selection is important even for deep learners [8,9]. Although deep learners gain good performance in image analysis, in other cases, differences between deep learners and other classification method are very small. It may be due to that we have not yet suitable network structure and we may not use suitable features for classification. The empirical fact that deep learners are good at recognition of images suggests that some kind of topological relations should be explicitly embedded into traini data. So, here we propose a new feature selection method based on correspondence analysis, which calculates mapping attributes to points of multi-dimensional coordinates. The method can embed the topological relations between keywords and concepts into the data.

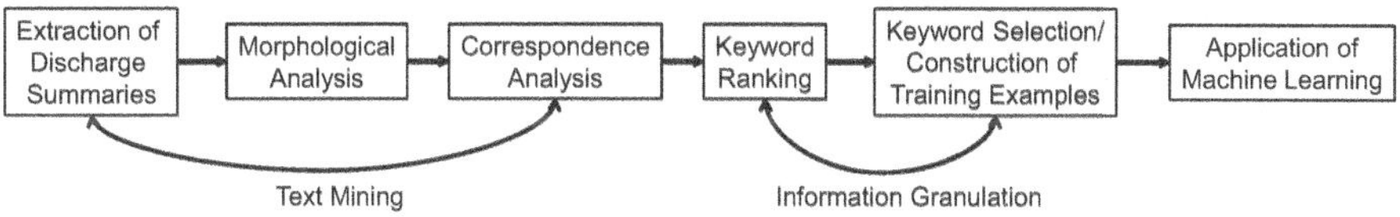

Fig. 1. Mining process pipeline

3.3 Mining Process

Figure 1 shows the proposed total mining process pipeline, composed of the following five steps. The first three methods can be called text mining process and the next two methods can be called information granulation process, which determines the granularity of knowledge for classification. Please note that these five methods are automated: after the users impose some conditions for text retrieval, such as collection period, and select the classification model, the process will start and the classification model will be generated.

Extraction of Discharge Summaries. First, discharge summaries are extracted from the hospital information system as texts.

Morphological Analysis. Next, morphological analysis (MeCab [2]), is applied, which outputs a term matrix, that is, a contingency table for keywords and concepts are generated.

Correspondence Analysis. Thirdly, correspondence analysis is applied to a term matrix. Although high dimensional coordinates can be selected, since a very large table is obtained, we focus on two dimensional analysis which can be easily used for visualization. To each key word and concept, two dimensional coordinate is assigned.

Ranking. The coordinates of a concept and a keyword will be used to calculate the euclidean distance between them. We use the distance values for ranking of keywords to each concept: the smaller the distance is, the higher its ranking is.

Keyword Selection. We assume the number of keywords selected before the analysis, say 100. Usually, since the lower rank keywords give information about specific cases, the number of selected keywords will determine the granularity of induced knowledge. Then, keywords whose ranking are up to 100 are selected keywords for classification. Some keywords may overlap, so such overlapped keywords will be deleted. Then, training examples with a classification label and the value of selected keywords (binary attributes) are constructed.

Classification. Finally, classification learning methods will be applied. In this paper, we selected random forest [6], deep learning (multi-layer perceptron) (darch), Support Vector Machine (SVM) [4], Backpropagation Neural Network (BNN) [11] and decision tree (rpart) [10] for comparison.

4 Experimental Evaluation

We selected top 20 frequent DPC codes in fiscal year of 2015 extracted discharge summaries from HIS in Shimane University Hospital. Table 1 shows their statistics and includes the averaged number of characters used in summaries.

Except for extraction from data from HIS, all the processes are implemented on R 3.5.0.

4.1 Mining Process

Correspondence Analysis. For morphological analysis, RMeCab [2] is used and the bag of keywords was generated. From the bag of keywords, a contingency table for these summaries are obtained. Then, correspondence analysis which is implemented in MASS package on R3.5.0 was applied to the table and two dimensional coordinate was assigned to each keyword and each class.[2] Figure 2 shows the two-dimensional plot of correspondence analysis.

[2] The method can also generate $p(p \geq 3)$-dimensional coordinates. However, higher dimensional coordinates did not give better performance that the experiments below.

Table 1. DPC top twenty diseases (fiscal year: 2015)

No	DPC	Cases
1	Cataract (lateral)	445
2	Cataract (bilateral)	152
3	Type II Diabetes Mellitus (except for keto-acidosis)	145
4	Lung Cancer (with surgical operation)	131
5	Uterus Cancer (without surgical operation) 121	
6	Lung Cancer (without surgical operation, chemotherapy)	
7	Uterus benign tumor	111
8	Lung Cancer (without surgical operation, with chemotherapy)	110
9	Shortage of Pregnancy	110
10	Injury of Elbow and Knee	99
11	Autoimmune Disease	96
12	Non-Hodgkin Disease	94
13	Pneumonia	86
14	Lung Tumor (without surgical operation nor chemotherapy)	85
15	Chronic Nephritis	83
16	Liver Cancer	82
17	Gallbladder Stone	82
18	Cerebral Infarction	80
19	Retinal Detachment	75
20	Fetal Abnormalities	75

Ranking. Next, the distances between the coordinate of a keyword and that of a class is calculated, and the ranking of keywords for each class was obtained. By using the ranking, a given number of keywords were selected to generate a classification table.

Keyword Selection. Table 2 shows the number of selected keywords for 10 and 20 Top DPC codes. In the case of 250 keywords, 250 keywords were selected for each DPC code and 5000 keywords in total. But, since the overlapped words should be removed, so only 1932 keywords were used for classification. Thus, some important keywords may be deleted due to the overlap if such words are frequently used at least in two diseases.

Table 3 shows selected top ten keywords for three top DPC codes. For comparison, the results by tf-idf are attached. Interestingly, keywords selected by

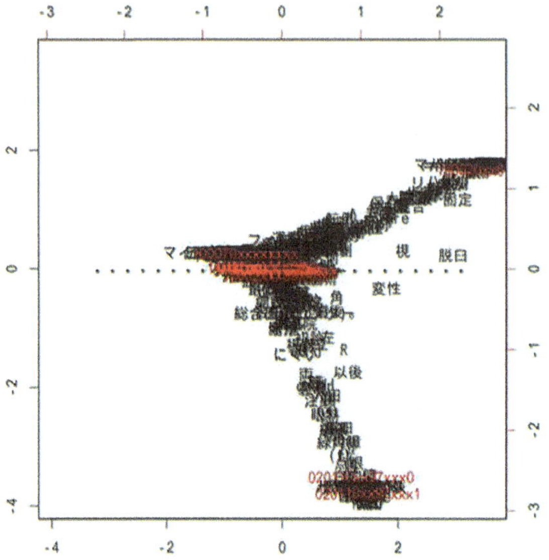

Fig. 2. Correspondence analysis

correspondence analysis are very different from ones selected by tf-idf, which suggests frequency based information may not play important roles in classification of discharge summaries.

Classification. Finally, decision tree (package: rpart [10]), random forest (package: randomForest [6], SVM (kernlab [4]), BNN(package: nnet [11]) and Deep Learner (multi-layer perceptron) (darch[3]) were applied to the generated table. For parameters of Darch, the number of intermediate neurons are 10, (10,5), (40,10) and (100,10), whose epoch was 100. For all other packages, the default settings of parameters were used.

Evaluation Process. Evaluation process was based on repeated 2-fold cross validation [5][4].

First, a given dataset was randomly split into training examples and test samples half in half. Then, training examples was used for construction of a classifier, and the derived classifier was evaluated by using remaining test samples. The above procedures were repeated for 100 times in this experiment, and the averaged accuracy was calculated.

---

[3] Darch was removed from R package. Please check the githb: https://github.com/maddin79/darch.

[4] The reason why 2-fold is selected is that the estimator of 2-fold cross-validation will give the lowest estimate of parameters, such as accuracy and the estimation of bias will be minimized.

Table 2. Number of selected keywords and actually used keywords

#Keyword	Selected keywords	
	DPC Top 10	DPC Top 20
1	10	19
2	19	37
3	27	54
4	36	71
5	44	88
10	88	167
20	115	309
30	247	449
40	334	597
50	406	718
100	724	1125
150	1000	1472
200	1192	1782
250	1382	1932
300	1547	2031
350	1676	2113
400	1797	2192
450	1929	2273
500	2028	2364
750	2304	2808
1000	2545	3000
ALL	13944	20417

Table 3. Selected Keywords for top three diseases (English translation)

Order	Cataract (Uni)		Cataract (Bil)		II-type diabetes	
	Corres	TF-IDF	Corres	TF-IDF	Corres	TF-IDF
1	in	IOL	eye	IOL	keton	mg
2	VA	PEA	(+),	PEA	CTR	ml
3	lower	sunbetazone	mature	(Dr)	hr	hour
4	DC	left	abastine	+	Enhance	g
5	Eye_drops	+	Vitreous	Cataract	FDP	blood_sugar
6	Vitreous	right	Allergy	execute	noise	
7	mature	disclose	lower	trouble	jumping_pain	
8	abastine	vegamox	In	visual Eyesight	hypotensive	
growth	Lung					
9	(+),	Cataract	VA	discharge	pancreatitis	admission
10	#	month	The	left	menopause	

The number of keywords varied from 1 to 1000, selected according to the rank given by correspondence analysis. For analysis, two units of HP Proliant ML110 Gen9 (Xeon E5-2640 v3.2 2.6 GHz 8Core, 64 GBDRAM) was used.

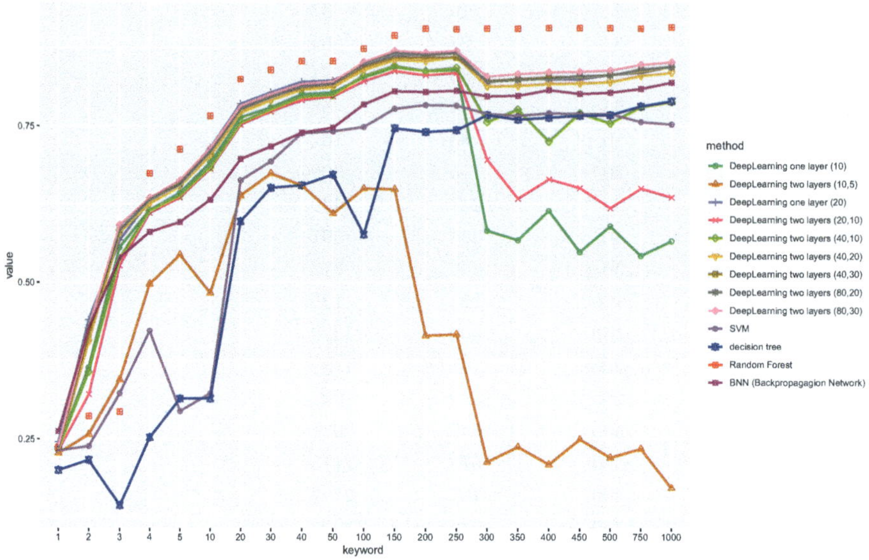

Fig. 3. Experimental results (Top 20)

4.2 Classification Results

Figure 3 plots the evaluation results for Top 20 diseases. The vertical axis denotes the averaged accuracy of a classifier, and the horizontal axis denotes the number of selected keywords.

Until the number of keywords is 4, all the classifiers gained around 70% accuracy. However, larger than 5, SVM was rather decreasing and other methods were monotonically increasing and became plateau with the number of keywords being 200. Random Forest achieved the best performance, compared with other classifiers. The second are Darch deep learners. It is notable that if more than 250 keywords were selected, the performance of Darch was getting worse. On the other hand, thoses of random forest and decision trees were monotonically increasing. Although the accuracy of BNN is worse than Darch (default setting) with 5 to 100 selected keywords, it approached that of Darch classifiers with larger number of keywords. Interestingly, the accuracy of decision tree was monotonically increased and the highest value was obtained when all the keywords were used for analysis.

5 Discussion

5.1 Misclassified Cases

Figure 4 and 5 show confusion matrices of random forest and darch (multi-layer perceptron), where DPC codes are set in order which means that similar diseases are assigned to similar codes.[5] Yellowed region indicates misclassified cases. It can be observed that whereas errors in random forest are located near the diagonal, errors in darch are more scattered. This suggests that random forest almost correctly classifies a case if similar dpc codes are grouped into one generalized class, while Darch has unexpected errors.

True \ Prediction		010060x099030x	020110xx97xxx0	020110xx97xxx1	020160xx97xxx0	040040xx97x0xx	040040xx9900xx	040040xx9904xx	040040xx99100x	040080x099x0xx	060050xx97x0xx	060340xx03x00x	070560xx99x0xx	100070xxxxxxxx	110280xx99000x	12002xxx99x40x	120060xx02xxxx	120180xx01xxxx	130030xx99x40x	140010x199x00x	160620xx01xxxx	Total
		brain infarction	Cataract (Uni)	Cataract (Bil)	Retina Detachment (Uni)	Lung Cancer (with Surgical Operation)	Lung Cancer (without Surgery)	Lung Cancer (with Bronchoscopy)	Lung Cancer (with Chemotheory)	Pneumonia	Liver Cancer (with Surgical Operation)	Goldbladder Stone & Inflammation	Severe Autoimmune Disease	Type II DB	Chronic Nephritis	Uterus Cancer	Uterus Benign Tumor	Fetal Abnormalities	Nonhodgikin	Pregnancy (early)	Injury of Knee or Elbow	
010060x099030x	brain infarction	17																				17
020110xx97xxx0	Cataract (Uni)		20	5																		25
020110xx97xxx1	Cataract (Bil)		33	34																		67
020160xx97xxx0	Retina Detachment (Uni)		1		28																	29
040040xx97x0xx	Lung Cancer (with Surgical Operation)					6		1		1												8
040040xx9900xx	Lung Cancer (without Surgery)						16	10	3	2					2							33
040040xx9904xx	Lung Cancer (with Bronchoscopy)						1	16														17
040040xx99100x	Lung Cancer (with Chemotheory)							2	29	1												32
040080x099x0xx	Pneumonia	1	3		1		1	1		17		2		1	1							28
060050xx97x0xx	Liver Cancer (with Surgical Operation)								1	1	11	4		1								18
060340xx03x00x	Goldbladder Stone & Inflammation		5								1	9										15
070560xx99x0xx	Severe Autoimmune Disease								2	1			13		1							17
100070xxxxxxxx	Type II DB										1			15								16
110280xx99000x	Chronic Nephritis	2	2							1			1	4	14							24
12002xxx99x40x	Uterus Cancer															6						6
120060xx02xxxx	Uterus Benign Tumor		1														4					5
120180xx01xxxx	Fetal Abnormalities																	7				7
130030xx99x40x	Nonhodgikin								1										21			22
140010x199x00x	Pregnancy (early)																			3		3
160620xx01xxxx	Injury of Knee or Elbow														1							1
	Total	20	65	39	29	6	18	30	36	24	13	15	14	21	19	6	4	7	21	3	0	390

Fig. 4. Confusion marix of random forest

5.2 Execution Time

Although the performance of BNN and Deep Learners were high, the problem is that they need more time for computation.

Figure 6 shows empirical comparison of repeated 2-fold cross validation (100 Trials). The vertical and horizontal axes denote computational speed and the number of keywords respectively. Whereas the time need for Random Forest

[5] DPC codes are three-level hierarchical system and each DPC code is defined as a tree. The first-level denotes the type of a disease, the second-level gives the primary selected therapy and the third-level shows the additional therapy. Thus, in the tables, characteristics of codes are used to represent similarities.

True \ Prediction		010060x099030x brain infarction	020110xx97xxx0 Cataract (Uni)	020110xx97xxx1 Cataract (Bil)	020160xx97xxx0 Retina Detachment (Uni)	040040xx97x0xx Lung Cancer (with Surgical Operation)	040040xx9900xx Lung Cancer (without Surgery)	040040xx9904xx Lung Cancer (with Broncho-scopy)	040040xx99100x Lung Cancer (with Chemotheory)	040080x099x0xx Pneumonia	060050xx97x0xx Liver Cancer (with Surgical Operation)	060340xx03x00x Gold-bladder Stone & Inflammation	070560xx99x0xx Severe Autoimmune Disease	100070xxxxxxxx Type II DB	110280xx99000x Chronic Nephritis	12002xxx99x40x Uterus Cancer	120060xx02xxxx Uterus Benign Tumor	120180xx01xxxx Fetal Abnormalities	130030xx99x40x Non-hodgikin
010060x099030x	brain infarction	13											1		3				
020110xx97xxx0	Cataract (Uni)		7	17								1							
020110xx97xxx1	Cataract (Bil)		25	41									1						
020160xx97xxx0	Retina Detachment (Uni)		2		27														
040040xx97x0xx	Lung Cancer (with Surgical Operation)					5		1	1		1								
040040xx9900xx	Lung Cancer (without Surgery)						24	5		2					1				
040040xx9904xx	Lung Cancer (with Bronchoscopy)						3	10	1							2			1
040040xx99100x	Lung Cancer (with Chemotheory)						3	2	26										1
040080x099x0xx	Pneumonia	2					2			11	2	1	7		3				
060050xx97x0xx	Liver Cancer (with Surgical Operation)			1			3				4	8		1					1
060340xx03x00x	Goldbladder Stone & Inflammation	2	2	1	1					2	2	5							
070560xx99x0xx	Severe Autoimmune Disease	1							2				6		8				
100070xxxxxxxx	Type II DB											1		13	2				
110280xx99000x	Chronic Nephritis	1	1				2			1	1			5	12			1	
12002xxx99x40x	Uterus Cancer			1			1					1				3			
120060xx02xxxx	Uterus Benign Tumor																5		
120180xx01xxxx	Fetal Abnormalities									2							2	3	
130030xx99x40x	Nonhodgikin					1					1								20
140010xx199x00x	Pregnancy (early)																		
160620xx01xxxx	Injury of Knee or Elbow																		
	Total	19	37	61	28	6	38	18	30	18	11	17	15	19	29	5	7	4	23

Fig. 5. Confusion matrix of deep learner

and SVM was 28 min in total, Deep Learners (10) needed 2 h 51 min (171min), Random Forest and BNN were close to Deep Learners. BNN was fast when the number of keywords was small, but the accuracy grew exponentially with more than 200 keywords. In the case of Darch, the number of intermediate layers costed high computational time, although the growth rate was smaller than BNN.

5.3 Comparison with tf-idf

A major approach in text classification is ranking with tf-ifd, which were introduced by Luhn [7] and Sparck Jones [3]. Thus, here, we compared tf-idf ranking with the above approach by using the same scheme as in Sect. 4.

Interestingly, deep learning with tf-idf ranking performs much worse than that with ranking by correspondence analysis, whereas random forest with tf-idf ranking is a little better than that with ranking by correspondence analysis when the number of selected keywords is smaller than 200. These results are clearly shown in Fig. 7, where the above results are superimposed. Thus, since deep learning achieves only a few percent differences in the averaged accuracy from random forest, ranking by correspondence analysis is a better approach for text classification by deep learning at least in this applied domain. It may be because ranking by correspondence analysis includes geometric information about keywords and concepts and such embedding of geometric knowledge is important for deep learning.

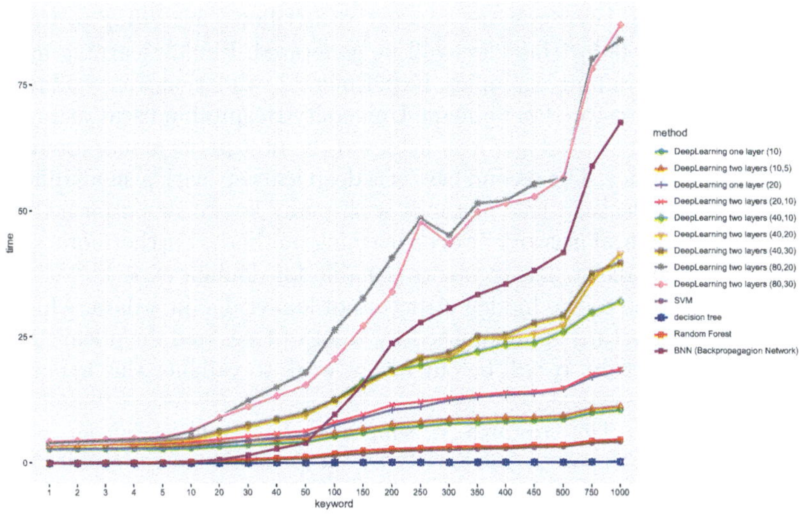

Fig. 6. Executed time for classification construction (Top 20)

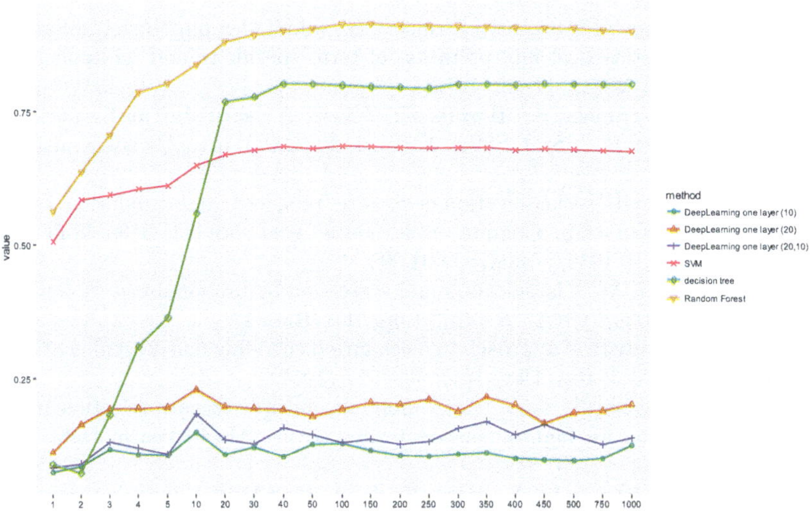

Fig. 7. Comparison of algorithms (tf-idf)

6 Conclusion

We propose a method for construction of classifiers for discharge summaries. First, morphological analysis is applied to a set of summaries and a term matrix is generated. Second, correspond analysis is applied to the classification labels and the term matrix and generates two dimensional coordinates. By measuring the distances between categories and the assigned points, ranking of key words will be generated. Then, keywords are selected as attributes according to the rank, and training example for classifiers will be generated. Finally learning methods are applied to the training examples. Experimental validation was conducted by discharge summaries stored in Shimane University Hospital in fiscal year of 2015. The results shows that random forest achieved the best performance around 93% classification accuracy. The second best was deep learners with a small difference, around 91%. But decision tree methods with many keywords performed only a little worse than neural network or deep learning methods. Furthermore ,selected keywords and the tree structure was reasonable for domain experts.

This may be due to the hidden structure of knowledge in a data, which may be close to the structure approximated by a set of trees and deep learning may generate such structure. It will be our future work to validate this hypothesis.

Acknowledgments. This research is supported by Grant-in-Aid for Scientific Research (B) 18H03289 from Japan Society for the Promotion of Science(JSPS).

References

1. Egakutsushinsha, Tokyo (2020)
2. Ishida, M.: Rmecab (2016). http://rmecab.jp/wiki/index.php?RMeCabFunctions
3. Jones, K.S.: A statistical interpretation of term specificity and its application in retrieval. J. Doc. **28**(1), 11–21 (1972)
4. Karatzoglou, A., Smola, A., Hornik, K., Zeileis, A.: kernlab - an S4 package for kernel methods in R. J. Stat. Softw. **11**(9), 1–20 (2004). http://www.jstatsoft.org/v11/i09/
5. Kim, J.H.: Estimating classification error rate: repeated cross-validation, repeated hold-out and bootstrap. Comput. Stat. Data Anal. **53**(11), 3735–3745 (2009). https://doi.org/10.1016/j.csda.2009.04.009
6. Liaw, A., Wiener, M.: Classification and regression by randomforest. R News **2**(3), 18–22 (2002). http://CRAN.R-project.org/doc/Rnews/
7. Luhn, H.P.: A statistical approach to mechanized encoding and searching of literary information. IBM J. Res. Dev. **1**(4), 309–317 (1957)
8. Mares, M.A., Wang, S., Guo, Y.: Combining multiple feature selection methods and deep learning for high-dimensional data. Trans. Mach. Learn. Data Min. **9**, 27–45 (2016)
9. Nezhad, M.Z., Zhu, D., Li, X., Yang, K., Levy, P.: SAFS: a deep feature selection approach for precision medicine. CoRR abs/1704.05960 (2017). http://arxiv.org/abs/1704.05960
10. Therneau, T.M., Atkinson, E.J.: An introduction to recursive partitioning using the RPART routines (2015). https://cran.r-project.org/web/packages/rpart/vignettes/longintro.pdf
11. Venables, W.N., Ripley, B.D.: Modern Applied Statistics with S, 4th edn. Springer, New York (2002). https://doi.org/10.1007/978-0-387-21706-2. http://www.stats.ox.ac.uk/pub/MASS4. iSBN 0-387-95457-0

Identifying Snowfall Clouds at Syowa Station, Antarctica via a Convolutional Neural Network

Kazue Suzuki[1(✉)], Masaki Shimomura[2], Kazuyuki Nakamura[2], Naohiko Hirasawa[3,5], Hironori Yabuki[3], Takashi Yamanouchi[3,5], and Terumasa Tokunaga[4]

[1] Hosei University, 3-7-2, Kajinocho, Koganei-shi, Tokyo, Japan
kazue.suzuki.88@hosei.ac.jp
[2] Meiji University, 4-21-1, Nakano, Nakano-ku, Tokyo, Japan
[3] National Institute of Polar Research, 10-3, Midori-cho, Tachikawa-shi, Tokyo, Japan
[4] Kyushu Institute of Technology, 680-4, Kawazu, Iizuka-shi, Fukuoka, Japan
[5] Graduate University for Advanced Studies (SOKENDAI), Shonan Village, Hayama, Kanagawa 240-0193, Japan

Abstract. This study evaluated the impact of Atmospheric River (AR) clouds on snowfall amounts based on limited observation data to estimate the surface mass balance (SMB) of Antarctica. To accomplish this, we attempted to identify the snowfall cloud at Syowa Station, Antarctica. We constructed a new convolutional neural network (CNN) architecture with multinomial and binary classifications for National Oceanic and Atmospheric Administration (NOAA) Advanced Very High Resolution Radiometer (AVHRR) images over five years. The CNN was based on VGG16, and concatenate layers were added as the inception module. We replaced all the convolution layers with global average pooling to reduce the number of parameters. Based on the positive CNN sample result, the multinomial classification emphasized the entire cloud structure, while the binary classification focused on cloud continuity. The results indicated accuracies of 71.00% and 65.37% for binary and multinomial classifications, respectively.

Keywords: CNN · Satellite image · Surface mass balance

1 Introduction

Recently, atmospheric rivers (ARs) have been observed with interest by the meteorological research community [7], being the cause of natural disasters around the world. ARs can bring a continuous moisture supply to specific regions, resulting in large amounts of rainfall and snowfall. In global climate change projections, the Antarctic ice sheet is known as one of the uncertainties in the hydrological cycle, ARs have been analyzed in several ways as an event affecting its water balance based on the climate modeling [6,12]. The main factor that has

K. Yada et al. (Eds.): JSAI 2020, AISC 1357, pp. 73–83, 2021.
https://doi.org/10.1007/978-3-030-73113-7_7

a positive effect on the water and mass balance of the Antarctic ice sheet is the phenomenon of snowfall. The net surface mass balance of the ice sheet due to snowfall has been analyzed by satellite and LIDAR observations, and seasonal and long-term variations have been reported [2,4,5]. This study evaluated the impact of AR clouds on amounts of snowfall based on the spatial synoptic patterns among certain elements (e.g., geopotential height, relative humidity, sea ice concentration) over several decades. Because snowfall amounts reported by various reanalysis data and regional climate models disagree considerably with the observed snowfall event amounts at Syowa Station [1], we must verify that the elements used in reanalysis data are sufficient to accurately interpret snowfall values.

To accomplish this goal, we investigated the relationship among atmospheric synoptic patterns and cloud patterns determined from satellite data, and defined these characteristic spatial patterns based on observation data obtained at Syowa Station (e.g., snow depth, weather condition, cloud amount). The patterns were applied as a training set for machine learning techniques to automatically identify similar patterns. By studying the influence on snowfall over Antarctica, we identified an AR cloud. In 2009, tremendous snowfall amounts were observed in East Antarctica (e.g., Princess Elisabeth Station, Syowa Station), and moisture transport was analyzed [3]. An AR in front of a cyclonic disturbance enhanced the poleward moisture flux outside Antarctica and carried significant snowfall into the region. Blizzard events at Syowa Station often exhibit similar atmospheric circulation conditions [8]. Based on these results, the present study focused on ARs as a means of learning and automatic identification via a convolutional neural network (CNN) in an effort to predict the snowfall in Antarctica.

2 Model and Data

2.1 Preliminary Study

We constructed an image classifier based on a CNN to apply to cloud images taken during the 2009 blizzard events. The image data were derived from the National Oceanic and Atmospheric Administration (NOAA) Advanced Very High Resolution Radiometer (AVHRR) channel 4. We added tags to the images based on the observed snowfall conditions. Clouds were identified in images tagged as positive, but no significant cloud was observed in images tagged negative. Based on Syowa Station snow depth data, we compared the spatial characteristics of the clouds with heavy versus light snow conditions. After several procedures, e.g., masks of Antarctic surface area, binarization, expansion processing, reduction of noise, we assume the boundary of the high-top clouds (HTC) area using the Chan-Vese Algorithm. HTC areas indicate where high levels of convection occur in the cloud. The cloud area (in number of pixels) corresponding to heavy snow differed by approximately one order of magnitude from the area corresponding to light snow (seen in Fig. 1). Although the CNN trial learning was overfitted due to the limited amount of sample data, we frequently observed satellite images with bright HTC regions, such as those found in ARs, during heavy snowfall.

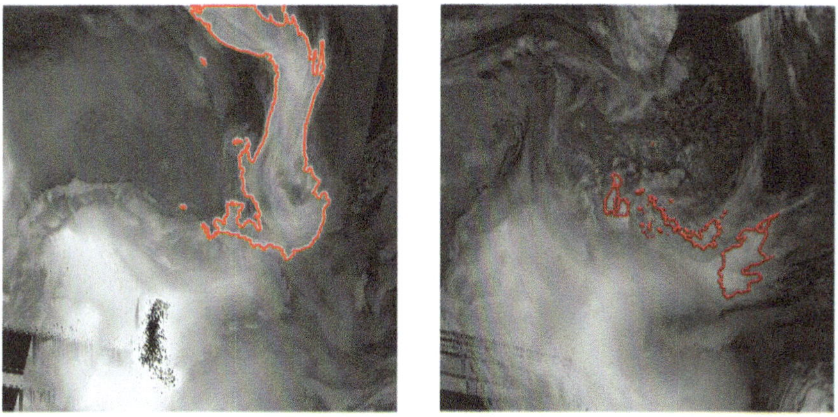

Fig. 1. AVHRR images of clouds taken during snowfall. Samples of HTC areas are outlined in red. Left: HTC for heavy snow on September 9th, 2009 and right: HTC for light snow on March 9th, 2009.

2.2 Model Description

To improve the accuracy of HTC identification, we constructed new CNN architectures as illustrated in Fig. 2 (multinomial classification) and Fig. 3 (binary classification). The new CNN was based on VGG16 [9] with three concatenate layers added as the inception module (first layer: 1 × 1 convolution; second layer: 1 × 1 convolution and 5 × 5 convolution; and third layer: 1 × 1 convolution, 3 × 3 convolution, and 3 × 3 convolution).

This study replaced all the convolution layers with global averaged pooling to reduce the number of parameters. We generated batch sizes of 16 and shuffled the images. A maximum of 300 epochs were used for training. The initial learning rate is 0.001, and it is multiplied by 10 each time the validation loss plateaus after 20 epochs. We employed cross-entropy as a loss function and Adam as an optimizer. The parameters were 0.9 for β_1 and 0.99 for β_2. The image files were 2400 × 2400 pixels, and there were 1531 images. Data augmentation method settings were: shift 1/255 from the original value in each pixel, add rotation at $\pm 10^{\circ}$ degrees randomly, change brightness [0.7, 1.0] randomly. Finally, we evaluated the CNN based on the accurate identification of the cloud.

2.3 Data Description

This study used NOAA AVHRR images (for the thermal infrared band) taken over five years. Due to the nature of polar orbit satellite observations, the field of vision in the images varied throughout each day. Initially, we combined several images to analyze the entire cloud structure. Based on the Syowa Station observations, clouds with associated snow weather conditions and cloud amounts over 8 (and up to 10, where 10 corresponds to complete cloud cover) were tagged as positive, as shown in Fig. 4.

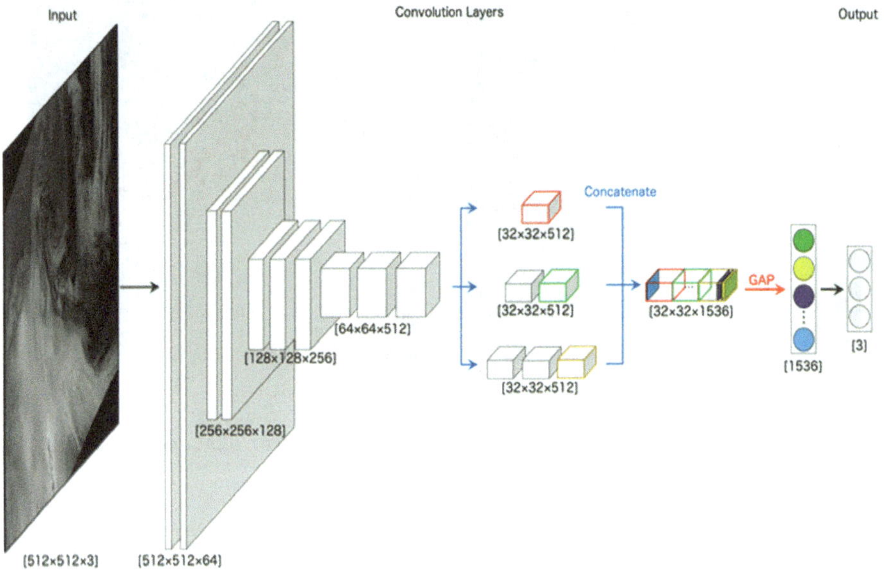

Fig. 2. A CNN architecture schematic, inception VGGNet for multinomial classification

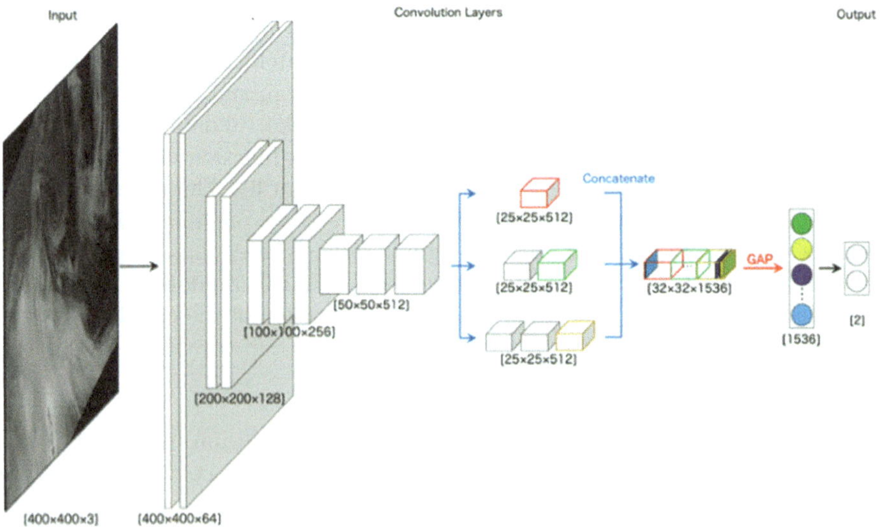

Fig. 3. A CNN architecture schematic, inception VGGNet for binary classification

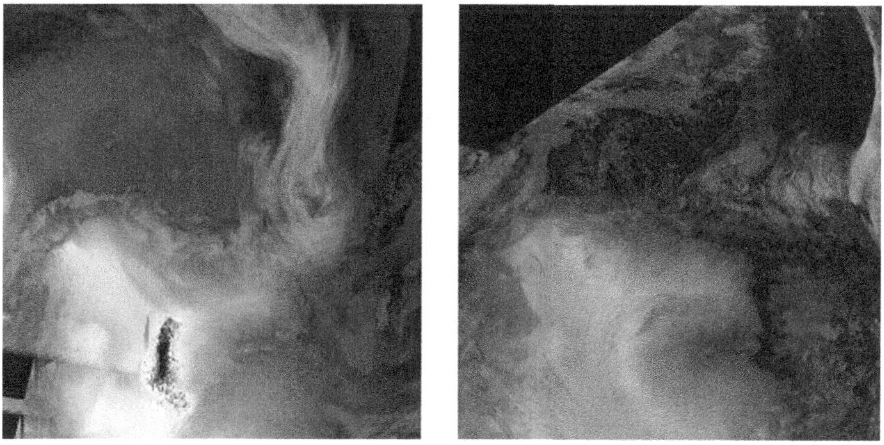

Fig. 4. Satellite images of snowfall at Syowa Station. Left: image that was tagged positive; right: image that was tagged negative.

The positive image shows a HTC in front of a cyclonic disturbance. Like trained cloud ARs, we hypothesize the HTC strongly influences the snowfall amount in Antarctica in this area. A semi-positive cloud is similar to a positive cloud, with the exception that it lacks snowfall at Syowa Station or includes only a partial cloud structure. Table 1 lists the numbers of images used for training, validation, and testing, categorized by their tags.

Table 1. Image sizes

	Training [80%]	Validation [5%]	Testing [15%]
Positive	138	6	26
Semi-positive	477	30	90
Negative	611	38	115

There is a clear bias between the positive and negative sizes, possibly due to lower learning accuracy in the multinomial classification. Binary classification attains nearly the same value in each situation.

2.4 Cropping Images

In this learning sample, we tagged (as positive) only snowfall clouds observed in the Syowa station snow event. However, the CNN can determine information regarding similarly patterned clouds with no snowfall at this station. To avoid this, we cropped images down to the yellow area shown in Fig. 5 surrounding Syowa and the sea for learning purposes, assuming that the cloud in this area affects the snowfall at Syowa station. A polarward moisture flux from mid latitude brought vapor transport to the station in [3]. In addition, the trajectories of air transport to Syowa Station during snowfall were calculated and that air percels assumed to carry high levels of water vapor primarily arrived at the station via this cropped ocean area, whereas they move along the coastline from their arrival to the station not in the snowfall conditions in [11].

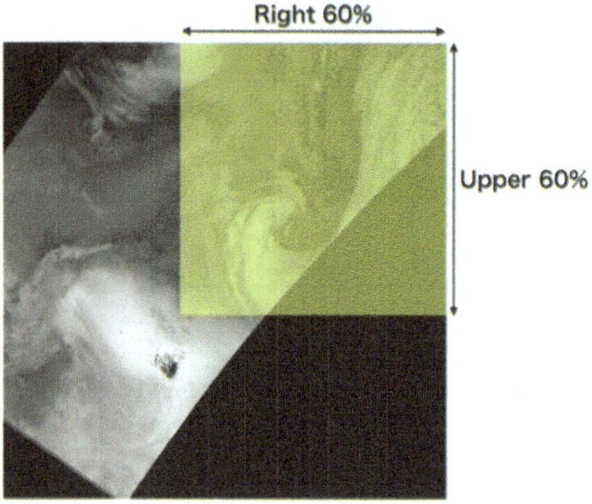

Fig. 5. Cropped satellite image

3 Results

For visual results from a large CNN class, we used Grad-CAM [10]. Figure 6 illustrates a positive sample case. The multinomial classification highlights the entire cloud structure, while the binary classification illustrates the cloud continuity.

In Fig. 7, the CNN captures the main cloud structure in the binary classification, and in the multinomial classification, notes cloud continuities and boundaries between the cloud thickness variations. Grad-CAM visualization suggests that the CNN can capture the entire structure of the HTC in multinomial classification.

Tables 2 and 3 define the results of multinomial and binary classifications, respectively. The accuracy of the multinomial classification is 65.37%, and of the binary classification is 71.00%. For the number of true positive (TP), true negative (TN), false positive (FP), and false negative (FN) values in the test data, as well as in the multinomial classification, we count whole numbers as the binary classification. While Grad-CAM visualized mapping implies that the multinomial classifier is extremely accurate, it is not superior to the binary classifier in terms of sensitivity and specificity. This may be due to the bias

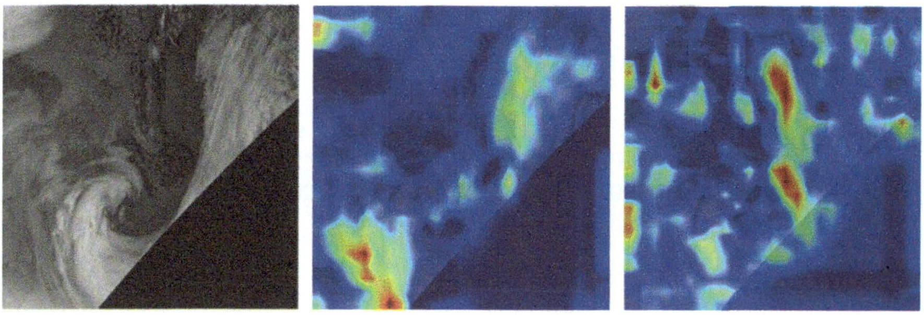

Fig. 6. Sample CNN identification results: cropped image (left), multinomial classification (center), binary classification (right).

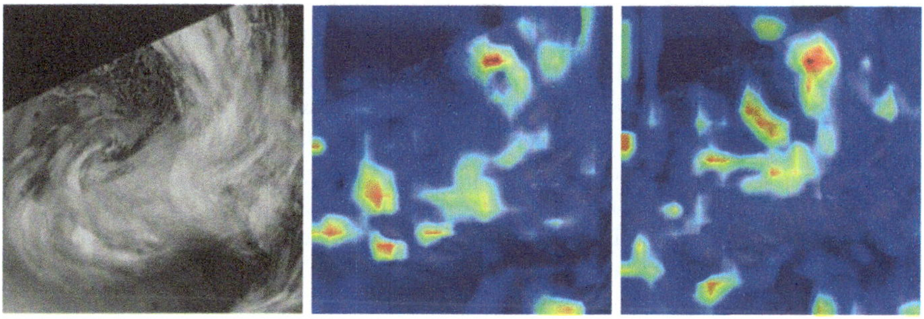

Fig. 7. Additional sample CNN identification results: cropped image (left), multinomial classification (center), binary classification (right)

of the sample size toward the multinomial. We added tags indicating whether the cloud was over Syowa Station. However, there was no improvement in the accuracy with either classification.

Table 2. Multinomial classification confusion matrix, sensitivity, and specificity

TP	TN	FP	FN
76	75	40	75
Accuracy	Recall	Precision	F-measure
0.6537	0.6552	0.6552	0.6252

4 Discussion

Before experimenting with InceptionVGGNet, we investigated the suitability of VGG16 and ResNet50, which are general-purpose CNN image classification

Table 3. Binary classification confusion matrix, sensitivity, and specificity

TP	TN	FP	FN
87	77	39	28
Accuracy	Recall	Precision	F-measure
0.7100	0.7565	0.6905	0.7220

models. The number of training data is generally on the order of thousands to tens of thousands when performing CNN image classification. However, the number of data in this study was 291 for positive, 408 for semi-positive, and 2091 for negative. The number of total images was relatively small, which led us to assume that overfitting would occur (in particular, there were fewer than 300 positive images). In fact, we performed binary classification with VGG16 and ResNet50, which predicted most of them as negative in the test (with accuracy of 53.7%). In addition, we checked the learning curve, and the validation error was not sufficiently improved. This also shows that our model is presumed to be overfitting.

There are two ways to improve the overfitting of CNN models: increasing the number of data and reducing the number of parameters. Because we could not expect an immediate increase in the number of data, we chose to simplify a model by reducing the number of parameters. The numbers of training parameters for both the InceptionVGGNet and the general-purpose CNN model used in this study are shown in Table 4. The input and output sizes are the same as those of inceptionVGGNet for binary classification (input: $400 \times 400 \times 3$, output: 2).

Table 4. Number of parameters for CNN model

CNN model	Trainable parameters
InceptionVGGNet	19,705,026
VGG16	33,655,618
ResNet50	23,591,690
2ConvModel	38,978

By including the Inception module, the weights of the convolutional layer can be made sparse, and large convolutions with a small number of parameters can be approximated [13]. In addition, while VGG16 has two fully connected layers after the convolutional layer, we installed a global average pooling layer in InceptionVGGNet. This is necessary to apply Grad-CAM, but it allows the omission of fully connected layers, which require many parameters. Consequently, the number of parameters is reduced by 41% compared to VGG16.

After these experiments, we gained additional image data for two years and made a new dataset. InceptionVGGNet did not identify the data well for the full seven year dataset. To investigate how InceptionVGGNet learns, we offered a simple CNN (2ConvModel) which has two convolutional layers for comparison (Fig. 8). Figure 9 denotes the visual explanation for positives in the multinomial classification, InceptionVGGNet was unable to catch anything whereas 2ConvModel was able to catch the cloud structure. From this result, the sample size in this case may not be enough to learn.

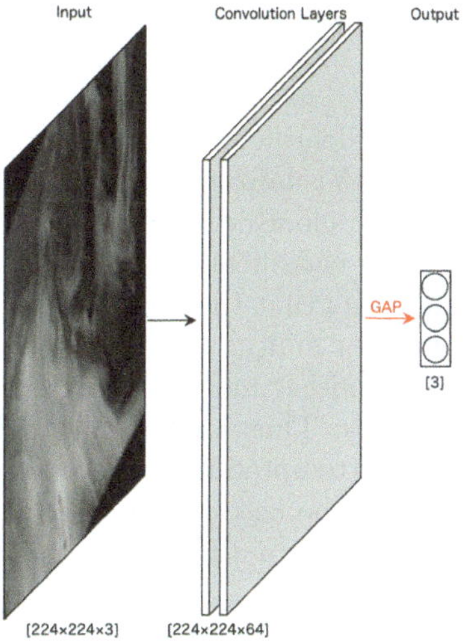

Fig. 8. CNN architecture, 2ConvModel

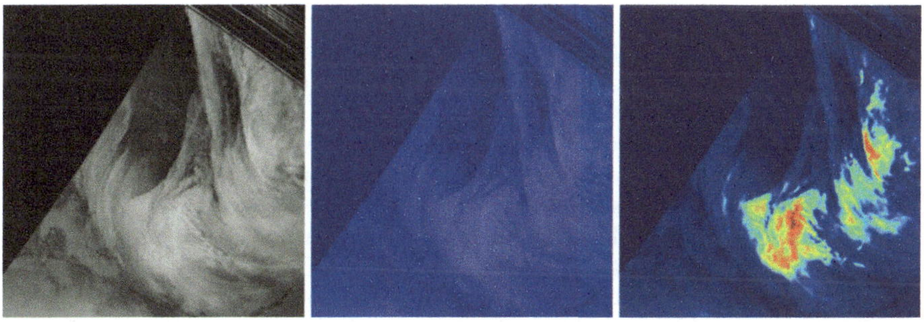

Fig. 9. Sample CNN identification results multinomial classification: cropped image (left), InceptionVGGNet (center), 2ConvModel (right)

We also calculated the number of parameters for 2ConvModel for comparison. The expected accuracy of this model was not promising due to its relative simplicity. However, considering the features captured by this model, the number of parameters in InceptionVGGNet is high. Therefore, building a simpler model and increasing the number of data (especially of positive cases) should be attempted to improve accuracy.

5 Summary

By studying clouds during heavy snowfall in 2009 blizzard events, we observed several AR HTCs, which carries moisture into Antarctica. Using NOAA AVHRR channel 4 images over five years, we developed a new CNN for HTC learning. Our results indicate that CNN identification is more accurate with binary classification than with multinomial. Visualization utilizing Grad-CAM suggests that the CNN focuses on the entire cloud structure for multinomial classification, and cloud continuity for binary classification. The undesirable level of accuracy achieved in this research suggests that there is a lack of adequate figures in the training data set, particularly for multinomial classification. Additional samples are required for learning and another method, e.g., U-Net, may be beneficial in capturing the AR cloud structure. Thus, building a simpler model and increasing the number of data should be attempted to improve accuracy. Further, the findings of this study can eventually be used to determine the surface mass balance of Antarctica, which we will pursue in a future study.

Acknowledgements. This work was supported by ROIS-DS-JOINT (009RP2019) and JSPS KAKENHI Grant Number 16K21585, 20K11718.

References

1. Agosta, C., et al.: Estimation of the Antarctic surface mass balance using the regional climate model MAR (1979–2015) and identification of dominant processes. Cryosphere **13**, 281–296 (2019)
2. Bromwich, D.H., et al.: Tropospheric clouds in Antarctica. Rev. Geophys. **50**, RG1004 (2012)
3. Gorodetskaya, I.V., Tsukernik, M., Claes, K., Ralph, M.F., Neff, W.D., Van Lipzig, N.P.M.: The role of atmospheric rivers in anomalous snow accumulation in East Antarctica. Geophys. Res. Lett. **41**, 6199–6206(2014)
4. Jolly, B., Kuma, P., McDonald, A., Parsons, S.: An analysis of the cloud environment over the Ross Sea and Ross Ice Shelf using CloudSat/CALIPSO satellite observations: the importance of synoptic forcing. Atmos. Chem. Phys. **18**, 9723–9739 (2018)
5. Kim, B.H., Seo, K.W., Eom, J., Chen, J., Wilson, C.R.: Antarctic ice mass variations from 1979 to 2017 driven by anomalous precipitation accumulation. Sci. Rep. **10**, 20366 (2020)
6. Ma, W., Chen, G., Guan, B.: Poleward shift of atmospheric rivers in the Southern Hemisphere in recent decades. Geophys. Res. Lett. **47**, e2020GL089934 (2020)
7. Payne, A.E., et al.: Responses and impacts of atmospheric rivers to climate change. Nat. Rev. Earth Environ. **1**, 143–157 (2020)

8. Sato, K., Hirasawa, N.: Statistics of Antarctic surface meteorology based on hourly data in 1957–2007 at Syowa Station. Polar Sci. **1**, 1–15 (2007)
9. Simonyan, K., Zisserman, A.: Very deep convolutional networks for large-scale image recognition, arXiv preprint (2014)
10. Selvaraju, R.R., Cogswell, M., Das, A., Vedantam, R., Parikh, D., Batra, D.: Grad-CAM: visual explanations from deep networks via gradient-based localization, arXiv preprint (2016)
11. Suzuki, K., Yamanouchi, T., Motoyama, H.: Moisture transport to Syowa and Dome Fuji stations in Antarctica. J. Geophys. Res. **113**, D24114 (2008)
12. Turner, J., et al.: The dominant role of extreme precipitation events in Antarctic snowfall variability. Geophys. Res. Lett. 3502–3511 (2019)
13. Uchida, Y., Yamashita, T.: Research trends in convolutional neural networks (in Japanese). The Institute of Electronics, Information and Communication Engineers, IEICE Technical Report, pp. 25–38 (2017)

Context-Aware Dialogue Response Generation Integrating Causal Relation

Bo Yang(✉), Jianming Wu, and Gen Hattori

KDDI Research, Inc., 2-1-15 Ohara, Fujimino-shi, Saitama, Japan
bo-yang@kddi-research.jp
https://www.kddi-research.jp/

Abstract. This is an extension from a selected paper from JSAI2020. In order to generate more content-rich responses, as well as avoid irrelevant or monotonous ones, some researchers set off to focus on dialogue generation models in which causal relation is also taken into consideration. However, it is hard to distinguish whether there is causal relation between speaker's and responder's utterances from dialogue automatically, which makes collecting such utterance pairs even harder. In this paper, we first propose a transfer learning method to learn causality features from web contents such as Wikipedia dataset and pre-train a causality classifier model. Subsequently, the causality classifier model is taken advantages into collecting utterance pairs containing causality, which is used in the re-ranking method of seq2seq generation model instead of rule-based one, to improve the dialogue diversity and context coherency. An evaluation is conducted to show that the proposed method improves diversity and coherency of dialogues generation from three aspects: 1. The causality classifier model used for detecting dialogue pairs is confirmed by three famous word embedding methods and two famous corpus. 2. The causality generation model improves response diversity, which is evaluated by automatic dialogue evaluation metrics. 3. The causality generation model improves dialogue context coherency, which is evaluated by human observes.

Keywords: Human-computer interaction · Dialogue generation · Neural network conversation model · Generative conversation model · Causal relation · Causality generation model

1 Introduction

Deep neural networks have achieved state-of-the-art performance in natural language processing (NLP) tasks, such as text classification, text comprehension, machine translation and dialogue generation. At almost the same time, [1] and [2] successively proposed a hierarchical recurrent encoder-decoder neural network to perform machine translations between two languages, which is now commonly referred to as the sequence-to-sequence (seq2seq) model. Following this, [3] applied the seq2seq model to address the dialogue generation problem.

K. Yada et al. (Eds.): JSAI 2020, AISC 1357, pp. 84–96, 2021.
https://doi.org/10.1007/978-3-030-73113-7_8

Compared to hand-written rules or manual annotation methodologies, neural conversational models (NCMs) demonstrated promising performance in terms of both effort-reduction and time-saving.

However, simple NCMs tend to produce monotonous or irrelevant responses [4], such as "Yes", "Sure", "I don't know", "I don't understand". To generate meaningful, logical and diverse responses, the seq2seq model was improved by introducing a hierarchical latent variable encoder-decoder structure in [5], facilitating both the generation of content-rich responses and the maintenance of a meaningful dialogue simultaneously. In [6], a non-task-oriented response generation model was constructed using causal knowledge collected by a set of response generation rules. As far as we know, it was the first attempt to study response generation integrating causal relation. Later, [7] proposed a re-ranking method, which re-ranked response candidates generated from conversational models by using causal events in a dialogue history and response candidates. These two methods certainly improved coherency and dialogue continuity of system responses to some extent. However, in both methods, a large causality corpus consisting of causal event pairs is required. Furthermore, the speaker's utterance along with the generated response candidate has to match exactly with a certain causal event pair, which is more like a rule-based method employed during a re-ranking process. Most importantly, the definition of causality in dialogue is limited to causal event pairs, but causality in sentences covers a wider range. The main contributions of this paper are: 1. As far as we know, this is the first exploration of a transfer learning method to learn causality from web content, which is used to detect causal features from dialogue pairs as a binary classification task. 2. The proposed causality generation model improves the diversity by integrating causality into the response and prevents it from generating monotonous replies, such as "Yes", "Sure". 3. The proposed causality generation model produces context-aware response, which maintains consistent causality between speaker's and responder's utterances, thus improves the context coherency of dialogue.

2 Related Works

2.1 Causality in Sentences

The task of causality detection has long been the focus of natural language processing (NLP) research. In most cases, however, causality detection is often viewed as a causal relation extraction task. The most common case involves detecting cause-effect event pairs such as "smoking causes cancer". However, as is discussed in [8], causal relation can be considered from the following aspects: 1. **Marked** or **Unmarked:** For example, "smoking causes cancer" is marked with the causality word "cause", in which "smoking" is cause and "cancer" is effect. However, "Bob became infected with the corona virus and was admitted to hospital", in which there is no specific linguistic unit such as "because" or "cause". 2. **Ambiguity:** For example, "Bob has been admitted to hospital because he became infected with the corona virus" uses ambiguous word "because". But, "Bob was admitted to hospital since he was infected" uses the unambiguous word "since". 3. **Explicit** or **Implicit:** For example, "Bob was attacked by a man and died" has explicit cause "be attacked" and effect "died". However, "Bob was killed" is implicit but includes cause "be killed" and effect "died".

2.2 Causal Sentence Detection

A transfer learning method was introduced in [9], which sheds light on detecting causal sentences, rather than extracting causal relation from sentences. In this paper, the model BIGRUATT consists of a bidirectional GRU (BIGRU) and self-attention and exploits transfer learning, namely ELMO [10] and BERT [11]. The authors claimed it was the first study to focus on causal sentence detection as a binary classification task. The evaluation shows that the proposed method significantly boosts causal sentence detection performance on small datasets such as SemEval [12], CausalTB [13], and EventSL [14].

2.3 Seq2seq Model and Beam Search

The basic architecture of seq2seq is a model with two parts: an encoder and decoder, which can be a Long Short-Term Memory (LSTM) module. Sentence $x = (x_1, x_2, ..., x_T)$ with words x_i is input into the encoder and $y = (y_1, y_2, ..., y_{T'})$ is its corresponding output from the decoder with words y_i. The dialogue generation task seeks to maximize the likelihood of $p(y|x)$

$$p(y|x) = \prod_{t=1}^{T'} p(y_t|x, y_1, ..., y_{t-1}) \tag{1}$$

In the decoding part, considering every possible y is intractable, thus, a heuristic search algorithm called the beam search method is usually used to generate the N-best candidate responses [15]. At each time step during the decoding phrase, N-best token extensions with the highest scoring extensions by a pre-defined scoring function are retained. This process is repeated until finally N-best candidate responses are obtained.

3 Proposed Methodology

3.1 Causality in Dialogue

Different from that in passage texts, there is hardly any causality verbs within speaker's and responsor's utterances, which makes collecting utterance turn with causal relation very difficult. Thus, we investigate a famous dialogue corpus, named Cornell Movie-Quotes Corpus [16], and mainly six kinds of causal relation are defined with examples [17], as is shown in Table 1. Some regular patterns or diversities can be dig out from the dialog data set as following: 1. Cause and effect is not limited to event pairs but can also be human action, phenomenon, state, psychological feature and so on. 2. Cause and effect is not always distributed, but

Table 1. Causal relation definition for dialogue

Causal type	Cause	Effect	Example
Causal query	Accident	Hurt hand	Speaker: How did you hurt your hand?
			Responsor: Just an accident. Nothing
Causal surmise	Globsssal warming	Bizarre weather	Speaker: A series of bizarre shifts in local weather patterns
			Responsor: Global warming?
Causal explication	Tension	Shake	Speaker: You're shaking.
			Responsor: It's the tension
Causal complement	Pride, integrity etc.	Not lick butt	Speaker: Because I had too much pride. Too much integrity. I wouldn't lick nobody's butt ...
			Responsor: Daddy, it can't be just because of that. There had to be other factors ...
Causal intervention	Not pay	Thumbs & fingers broken	Speaker: What if I don't pay ya, Bert?
			Responsor: You don't pay me? You gonna get your thumbs broken. And your fingers.
Causal counterfactual	High voice	Scare son	Speaker: Will you please keep your voice down, there's people ...
			Responsor: You could have scared the son of a bitch off

in some cases only exits in either speaker's or responsor's utterance. 3. The same as considering causal inference, it can be investigated from three layer causal hierarchy:

- **Association:** One speaker asks for the cause or effect by queries like "What is the reason?", "why?". Causal query, causal surmise, causal explication and causal complement are all belongs to this layer.
- **Intervention:** Start query using "what if?" or "what if I do?". Causal intervention is general in dialogue and it needs inference based on existing causal knowledge.
- **Counterfactuals:** Using query like "what if I had not?". It usually calls for imagining of human being for causal inference rather than existing causal knowledge.

4. There is no need for causality in dialogue always to be correct, and response can be speculation, doubt, gainsay and so on.

5. In some cases, causality will not be easily discovered unless context information is taken into consideration.

Based on these constraints, the causal relation of dialogue is becoming more specific. Unlike [6] and [18], in which dialogue's causality is merely limited to casual relation pairs existing in speaker's and responsor's utterance separately, and research scope will become wider.

It is stressed in [19] that almost 33.3% of dialogue utterances contains causal relation, thus causality is also playing quite an important role in dialogues. The most common method for automatic extraction of causal relations from natural language texts is described in [20], by utilizing ambiguous causation verbs, which is also known as causal relation cues, namely "arouse", "lead to", "cause" and so on. Yet, as is explained that there is hardly any causation verbs between speaker's and responsor's utterances, which means collecting utterance turn containing causality using causality cue words almost impossible.

Since our goal is to collect causal dialogue as many as possible for dialogue generation but not relation extraction, we can solve the problem from a new perspective: Do as human being, learn causality feature from materials such as books, magazines, newspapers, or Internet, then use it to pick up causality utterance turn from dialogue. And dialogue generation with causality will make the response more rich in meaning and full of logic rather than monotonous ones.

3.2 Causality Transfer Learning Model

As is discussed in [19] that almost 33.3% of dialogue utterances contain causal relation, which supports our assumption that causality plays a significant role in conversation. In the automatic causal relation extraction task, many ambiguous causation verbs are utilized, such as "arouse", "cause", and "lead to". However, there are hardly any causation verbs between the speaker's and responder's utterances, which makes automatic detection of causal relation by causality cues almost impossible. Yet, our goal is not automatic extraction but to collect speaker and responder utterance pairs containing causality, which is used in the dialogue generation model to produce responses and make them rich in meaning and fully logical, rather than a succession of monotonous responses. Referring to [21], thirty causal verbs such as "lead to", "cause", "effect of" are utilize to construct a causality verbs corpus. Then we propose a causality transfer learning model as is described in follow steps:

Step 1: Wikipedia dump data is collected from the website periodically, and sentences containing causality verbs are filtered out as positive data, while, the same number of sentences without causality verbs in the same passage are sampled as negative data. In this way, the amounts of positive data and negative data are designed to be almost the same size, which can make the neural model converge quickly.
Step 2: It should be noted that causality verbs must be deleted to reduce the impact of such verbs on classification results.
Step 3: positive data and negative data are shuffled and split into training/validation/test data accordingly.
Step 4: the causality classification model is trained, this will be introduced later. When using the model to predict causality dialogue pairs, just join utterance turn to create a single sentence and input it into the model.
Step 5: the causality dialogue pairs are collected into causality dialogue data as one part of training data for the dialogue generation model.

3.3 Causality Classification Model

However, different from that in passage texts, there are hardly any causality verbs within the speaker's and responder's utterances, which makes collecting utterance turn with causal relation very difficult. Some regular patterns or diversities can be extracted from the dialogue dataset as followings: 1. Cause and effect is not limited to event pairs but can also be human actions, phenomena, states, psychological features and so on. 2. Cause and effect is not always distributed, but in some cases only exists in either the speaker's or responder's utterance. 3. There is no need for causality in dialogue always to be correct, and a response can express speculation, doubt, denial and so forth. 4. In some cases, causality will not be easily discovered unless contextual information is factored in consideration. Based on these constraints, the scope of causal relation is becoming wider but not limited to casual event pairs. As is elucidated by [9], we propose a causality classification model that focuses on detecting causality sentences, rather than extracting causal event pairs from sentences. Our final goal is to generate dialogue responses containing causality with a speaker's utterance. The benefit is apparent in that we do not have to prepare causal event pairs (cause and effect) at the vary beginning. Suppose we have a sentence with words $(w_1, w_2, ..., w_T)$, and the sentence is embedded with embedding layer to be $x_i = W_e w_i$ with embedding matrix W_e. Then bidirectional GRUs [1] is exploited to get information from both directions from a certain sentence, which is defined as hidden state h_i by concatenating $\overrightarrow{h_i}$ and $\overleftarrow{h_i}$ with $h_i = [\overrightarrow{h_i}, \overleftarrow{h_i}]$,$i \in [1, T]$. The overall formula is as follows

$$h_i = [\overrightarrow{h_i}, \overleftarrow{h_i}] = [\overrightarrow{GRU}(W_e w_i), \overleftarrow{GRU}(W_e w_i)] \tag{2}$$

Since we can obtain a representation of the sentence's meaning, the most important thing is to know which word has the most impact on the final classification of causality. Inspired by [22], wherein it was described how a words in sentences containing causality, we also take advantage by adding a word level attention. Hidden state h_i is input into an Multi-Layer Perception (MLP) layer to get a hidden representation u_i. It is then evaluated with a word level context vector u_w and a normalized importance weight α_i through a softmax function. Then α_i is viewed as a word weight and weighted sum based on h_i, is calculated to obtain the final sentence vector v. The overall word attention formula is as follows

$$u_i = tanh(W_w h_i + b_w) \tag{3}$$

$$\alpha_i = \frac{exp({u_i}^T u_w)}{\sum_T exp({u_i}^T u_w)} \tag{4}$$

$$v = \sum_t \alpha_i h_i \tag{5}$$

Finally, a softmax function is applied to the sentence feature v and the negative log likelihood of label (1 for a causality sentence and 0 for a non-causality sentence) is used as training loss.

3.4 Causal Re-Ranking Model

In response to the speaker, in order to make the neural conversation network generate as many responses containing causality as possible, a causality re-ranking model is proposed. A detailed overview of the model is shown in Fig. 1. Dialogue data consisting of dialogue utterance pairs is split and shuffled into training/validation/test dataset and used to train a Seq2Seq model with global attention. In the decoding process, whenever a sentence s_i from dialogue test data is input into the Seq2Seq model, N-best responses $(r_1, r_2, ..., r_N)$ with sentence r_i can be obtained when a beam search is implemented. Then N-best sentences are formed by certain test utterances and N-best responses to be $(s_i r_1, s_i r_2, ..., s_i r_N)$. We input the N-best sentences into a causality classifier one by one and calculate the probability of causality. If the probability is higher than a pre-defined threshold, that vary sentence $s_i r_i$ is re-ranked to the first place. Finally, the sentence with the highest probability is selected and the response it contains is chosen as the final output.

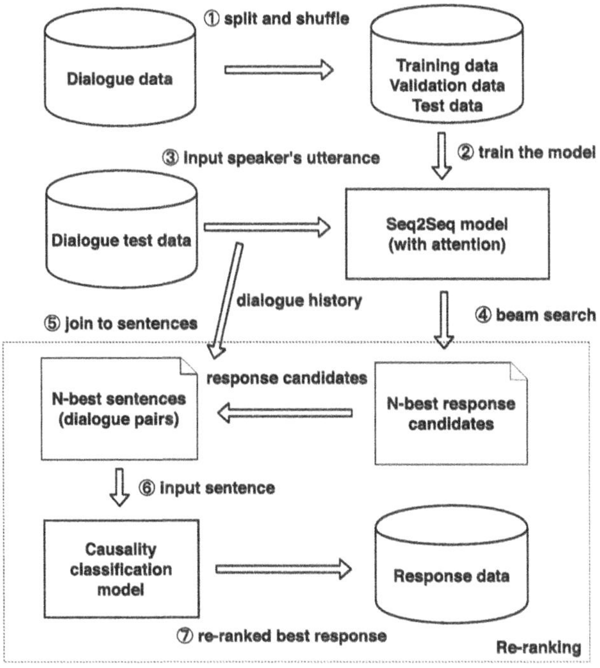

Fig. 1. Causal re-ranking model

4 Evaluation

4.1 Evaluation 1: Causality Classification Model

Evaluation Setup. From 3,000,000 English sentences in Wikipedia dump data, 64658 sentences containing causal verbs are extracted in order to learn more causality features. Thirty causal verbs are described in proposed methodology. All causal verbs in the sentences are deleted to diminish their effect. Sentences including causality verbs are marked as '1' while the sentences without causality verbs are labeled with '0'. All data along with labels are firstly shuffled and then spit into training data, validation data, and test data in a ratio of 7:2:1. Transfer learning is implemented by using one of Word2Vec [23], BERT [11] or GPT2 [24] as the embedder separately. Entropy loss is used for training of the model. During training, batch size is set to 64 and the best learning rate is searched using a grid search on the validation data. The maximum length of an input sentence is limited to 20 words. To test the performance of the pre-trained model in classifying dialogue data, the Cornell Movie-Dialogue Corpus from [16] is used. The corpus contains 220,579 conversational exchanges between 10,292 pairs of movie characters, with a total of 304,713 utterances. From the corpus, 300 causality utterance pairs and 300 non-causality utterances are randomly collected according to the causality defined by [8].

Table 2. Causality classification model results

Embedder	Test set	Precision	Recall	F1-score
Word2Vec	Wiki	88.22	89.45	88.83
	Dialogue	62.12	64.34	63.21
BERT	Wiki	96.01	96.80	96.40
	Dialogue	**68.17**	**72.44**	**70.24**
GPT-2	Wiki	92.46	90.01	91.22
	Dialogue	65.41	68.25	66.80

Results and Analysis. The evaluation results are shown in Table 2. For each embedder and data set, the evaluation is repeated 10 times and the average value is recorded as the final result. Three metrics: precision, recall, and F1-score are taken into consideration. Our task is to learn causality from web content using learned features to classify dialogue pairs; this is a new task for which there is no comparative benchmark. Thus, we use three kinds of embedder to ascertain which one is the most appropriate. It is apparent that the BERT embedder achieved better results when used on the dialogue data set compared with the other two. Due to the feature differences between dialogue data and Wikipedia data, the classification results also show an apparent difference. The reasons for this difference can be attributed to the following factors:

1. The design of the model, which can be improved and attains a higher accuracy.
2. The sentence containing causal words does not guarantee the causality of the sentence.
3. The common vocabulary for a different domain does not share the same distribution.

From detected results, causal dialogue pairs such as ("She hates smokers.", "It's a lung cancer issue.") from the Cornell Movie-Dialogue Corpus can be found out by existing methods by searching cause-effect pair (lung cancer issue, hate smokers). However, we can also find some examples without a cause-effect pair that cannot be detected using existing methods, such as ("Why?", "Because she'll scare them away."), and examples with cause and effect in either speaker's or responder's utterances, such as ("Are you crying.", "She's crying because she's so happy").

4.2 Evaluation 2: Diversity Evaluation of Causality Re-Ranking Model

Evaluation Setup. All the dialogue pairs data in the Cornell Movie-Dialogue Corpus are used as the dialogue data. It is split and shuffled into training data, validation data, and test data in a ratio of 8:1:1. Training data are used for training the Seq2Seq model, validation data are also used in the training process to avoid overfitting, and test data are for evaluation of the model. The benchmark is the Seq2Seq with global attention, in which encoder and decoder are bidirectional GRUs. As the baseline, a greedy decoding algorithm is used, while beam search algorithm is implemented for the proposed method. In the training process, the maximum sequence length of an utterance in the Cornell Movie-Dialogue Corpus is set to 10, which means the maximum number of joined sentences with a Cornell utterance pair is 20. An ADAM optimizer [25] is used and the learning rate is set to 0.0005. The training process is designed to minimize the cross entropy loss, which is also the logarithm of perplexity. Finally, in re-ranking process, the threshold of causality is set to 0.5.

Table 3. Causal re-ranking model diversity evaluation

	Distinct-1	Distinct-2	BLUE
Baseline	0.033	0.121	0.011
1-best	0.032	0.120	0.012
2-best	0.034	0.122	0.012
3-best	**0.038**	**0.141**	**0.013**

Results and Analysis. As is well known, it is hard to evaluate the causality of dialogue pairs. However, to generate a response that recognizes causality in the initial speaker's utterance also means that a diverse range of responses can be produced. Thus, three metrics are used to evaluate the performance of the causal re-ranking model: distinct-1, distinct-2 and BLUE. Distinct-N [26], most notably distinct-1 and distinct-2, is a metric that measures the diversity of a sentence. It focuses on the number of distinct n-grams of a sentence and thus penalizes sentences with lots of repeated words. BLUE [27] is used to evaluate the machine translation: the higher the blue's score is, the more accurate the translated sentence is. But for dialogue generation, it is hard to tell whether a response is right or wrong, which also means considering BLUE on its own makes no sense. So BLUE should be considered along with a distinct metric to make the dialogue generation model produce responses full of diversity. In this paper, distinct-1 and distinct-2 count the number of distinct unigrams and bigrams, accordingly. BLUE is computed using 1-gram to 4-grams.

The evaluation results of the causal re-ranking model are shown in Table 3. The proposed method is evaluated with 1-best, 2-best, and 3-best. Compared with the baseline, they all make improvement on the score of BLUE, distinct-1, and distinct-2 to some extent. For distinct-1, compared with baseline, 1-best, 2-best, 3-best increased the score by −3.12%, 3.12% and 12.12%, respectively. For distinct-2, compared with baseline, 1-best, 2-best, 3-best increased the score by -0.8%, 0.8%, 16.53%, respectively. The trend is that from 1-best to 3-best, the diversity of generated responses is improved accordingly. And the amount of average token in 3-best is the largest. For BLUE, compared with baseline, 1-best, 2-best, and 3-best increased the score by 8.33%, 8.33%, 18.18%, respectively. To conclude, the results showed that the proposed method increased the diversity of responses.

4.3 Evaluation 3: Coherency Evaluation of Causality Re-Ranking Model

Evaluation Setup. Because there is no existing method able to evaluate the causality of dialogue pairs, it is performed by relying on human judgement. From the dialogue test data, 1000 sentences are randomly selected as speaker's input. One thousand sentences are input into a baseline model and the proposed model respectively, and the output is 1000 responses. Input and output are matched to make 1000 dialogue pairs. Five persons are asked to evaluate these pairs based on the following standards: Class 1. if there is explicit causal relation; Class 2, if there is implicit causal relation; Class 3, if there is week causal relation; Class 4, if there is no causal relation.

Table 4. Causal re-ranking model coherency evaluation

	1	2	3	4
Baseline	4.1%	5.3%	8.2%	82.4%
1-best	3.8%	7.1%	8.8%	80.3%
2-best	5.3%	8.5%	12.5%	73.7%
3-best	8.5%	10%	15%	63.5%

Results and Analysis. The evaluation results are shown as in Table 4. On the one hand, compared with baseline, 1-best, 2-best, and 3-best have increased 1, 2, 3 for classes 1, 2, 3. On another hand, class 4 decreased. In particular, 3-best showed the greatest increase in causality because more candidate responses are generated, meanwhile the no-causality proportion has decreased accordingly. Referring to [7] and [26], the results concluded that context coherency was increased in dialogue by integrating causality.

5 Discussion

Although the proposed method is better than existing methods [6,7] in terms of time-saving and versatility, a large causal relation pairs corpus is not available, so that the proposed method cannot be compared with existing methods from a performance perspective. Not only in the causal pairs collecting stage various format of causal dialogue pairs can be found compared with existing methods, but also the diversity of causal dialogue can be obtained in the generation stage. The previous work can only produce responses as following: ("I am very stressed.", "You should take a rest.") by referring to a causal event pairs corpus (stress, take rest). The proposed method can also produce responses like baseline by utilizing a deep-learning classifier, which is an automatic approach rather than a rule-based one. Furthermore, the proposed method can also produce a response containing both effect and result, rather than distributed in dialogue pairs: ("I am not feeling good.", "you are tired, it would be advisable to take a rest"). Last but not least, we have not considered the causal level of dialogue pairs, which is an issue that needs to be addressed in the future.

6 Conclusions

In this paper, a transfer learning based causality classification model is introduced to detect causal dialogue pairs, the usability of which is confirm by three famous embedders and two datasets. Then a causal re-ranking response generation model is introduced to generate dialogue responses that take both context information and causality into account. The results show that the proposed method can improve the diversity and coherency of responses compared with baseline that only use the seq2seq with attention from aspects of automatic evaluation metrics and human judgement. In the future, we are going to develop a causality evaluation metric, which can be used to evaluate the causal level of dialogue pairs.

References

1. Cho, K., et al.: Learning phrase representations using RNN encoder-decoder for statistical machine translation. In: Proceedings of the Empirical Methods in Natural Language Proceedings (EMNLP) Conference (2014)
2. Sutskever, I., Vinyals, O., Le, Q.V.: Sequence to sequence learning with neural networks. In: Neural Information Processing Systems (NeurlPS) Conference (2014)
3. Serban, I.V., Sordoni, A., Bengio, Y., Courville, A., Pineau, J.: Building end-to-end dialogue systems using generative hierarchical neural network models. In: Proceedings of the 30th AAAI Conference on Artificial Intelligence (AAAI-16) (2016)
4. Colombo, P., Witon, W., Modi, A., Kennedy, J., Kapadia, M.: Affect-driven dialog generation. CoRR (2019)
5. Serban, I.V., et al.: A hierarchical latent variable encoder-decoder model for generating dialogues. In: Proceedings of the 31th AAAI Conference on Artificial Intelligence (AAAI-17) (2017)
6. Sato, S., Inoue, N., Inui, K., Higuchi, S., Watanabe, N.: Towards generating non-task-oriented conversational responses using causal knowledge base. In: The 30th Annual Conference of the Japanese Society for Artificial Intelligence (JSAI2016) (2016)
7. Tanaka, S., Yoshino, K., Sudoh,K., Nakamura, S.: Conversational response re-ranking based on event causality and role factored tensor event embedding. In: 1st Workshop NLP for Conversational AI (ConvAI), ACL 2019 Workshop (2019)
8. Blanco, E., Castell, N., Moldovan, D.: Causal relation extraction. In: Proceedings of the International Conference on Language Resources and Evaluation (2008)
9. Kyriakakis, M., Androutsopoulos, I., Ametlle, J.G., Saudabaye, A.: Transfer learning for causal sentence detection. In: Proceedings of the 18th BioNLP Workshop and Shared Task (2019)
10. Peters, M.E., et al.: Deep contextualized word representations. In: Proceeding of the Conference of the NAACL: Human Language Technologies (2018)
11. Devlin, J., Chang, M.-W., Lee, K., Toutanova, K.: BERT: pre-training of deep bidirectional transformers for language understanding. arXiv preprint arXiv:1810.04805 (2018)
12. Hendricks, I., et al.: SemEval-2010 task 8: multi-way classification of semantic relations between pairs of nominals. In: Proceedings of the Workshop on Semantic Evaluations: Recent Achievements and Future Directions (2018)
13. Mirza, P., Sprugnoli, R., Tonelli, S., Speranza, M.: Annotating causality in the TempEval-3 corpus. In: Proceeding of the EACL Workshop on Computational Approaches to Causality in Language (2014)
14. Caselli, T., Vossen, P.: The event storyline corpus: A new benchmark for causal and temporal relation extraction. In: Proceedings of the Events and Stories in the News Workshop (2014)
15. Li, J., Monroe, W., Jurafsky, D.: A simple, fast diverse decoding algorithm for neural conversation models. arXiv preprint arXiv:1611.08562 (2016)

16. Danescu-Niculescu-Mizil, C., Lee, L.: Chameleons in imagined conversations: a new approach to understanding coordination of linguistic style in dialogs. In: Proceedings of the Workshop on Cognitive Modeling and Computational Linguistics. ACL (2011)
17. Yang, B., Wu, J., Hattori, G.: A transfer learning method of data collection for dialogue response generation concerning causal relation. In: The 34th Annual Conference of the Japanese Society for Artificial Intelligence (JSAI2020) (2020)
18. Tanaka, S., Yoshino, K., Sudoh, K., Nakamura, S.: Conversational response re-ranking based on event causality and role factored tensor event embedding. In: Proceeding of the First Workshop on NLP for Conversational AI (2019)
19. Ryoko, T., Ryuta, T.: An analysis of 'distinctive' utterances in non-task-oriented conversational dialogue. Trans. Jpn. Soc. Artif. Intell. **22**, 425–435 (2007)
20. Asghar, N.: automatic extraction of causal relations from natural language texts: a comprehensive survey. In: Proceeding of the 19th Annual SIGdial Meeting on Discourse and Dialogue (2016)
21. Luo, Z., Sha, Y., Zhu, K.Q., Hwang, S., Wang, Z.: Commonsense causal reasoning between short texts. In: Proceedings of Fifteenth International Conference on Principles of Knowledge Representation and Reasoning (KR2016) (2016)
22. Yang, Z., Yang, D., Dyer, C., He, X., Smola, A., Hovy, E.: Hierarchical attention networks for document classification. In: Proceedings of the 2016 Conference of the North American Chapter of the Association for Computational Linguistics: Human Language Technologies (2016)
23. Mikolov, T., Chen, K., Corrado, G., Dean, J.: Efficient estimation of word representations in vector space. arXiv:1301.3781 (2013)
24. Radford, A., Wu, J., Child, R., Luan, D., Amodei, D., Sutskever, I.: Language models are unsupervised multitask learners. OpenAI **1**, 9 (2019)
25. Kingma, D., Ba, J.: Adam: a method for stochastic optimization. In: International Conference on Learning Representations (2014)
26. Li, J., Galley, M., Brockett, C., Gao, J., Dolan, B.: A diversity-promoting objective function for neural conversation models. In: Proceedings of the 2016 Conference of the North American Chapter of the Association for Computational Linguistics: Human Language Technologies (2016)
27. Ram, A., et al.: Conversational AI: the science behind the alexa prize. In: 1st Proceedings of Alexa Prize (Alexa Prize 2017) (2017)

Impact of Domain Knowledge Quality on Inverse Reinforcement Learning

Reed Sogabe[1], Dinesh Bahadur Malla[2,3], Masaru Sogabe[3], Kitsuyoshi Sakamoto[2], and Tomah Sogabe[2,3,4](✉)

[1] K International School Tokyo, Tokyo, Japan
[2] Engineering Department, University of Electro-Communications, Chofu, Japan
sogabe@uec.ac.jp
[3] Grid-Inc, Tokyo, Japan
[4] i-PERC, The University of Electro-Communications, Chofu, Japan

Abstract. Incorporating domain knowledge into conventional reinforcement learning has proven to be difficult due to its inability to fully extract the features of the demonstrator. We thus propose the use two algorithms for inverse reinforcement learning, Bayesian Neural Network and Maximum Entropy to deal with this issue. The primary objective of this work is to determine if varying qualities of domain knowledge, in the form of a demonstrator, would have any significant impact on the rewards obtained from the two algorithms by applying it to the mountain car environment.

Keywords: Domain knowledge · Inverse Reinforcement Learning · Bayesian Neural Network

1 Introduction

Conventional reinforcement learning is based primarily on the idea that the learning agent learns and decides the next action, based on a predetermined policy. In reinforcement learning, the agent must learn a policy to maximize the expected returns R_t over a finite horizon T within which the agent updates and optimizes a discounted reward $\gamma \cdot r_t$ over time. In most of the cases, the policy π is defined by the mathematical simulation of the state action value function $Q(s_t, a_t)$ shown in Eq. (1).

$$Q^{\pi}(s_t, a_t) = \mathbb{E}_{r_t, s_t, \sim E,\ a_t \sim \pi}[R_t | s_t,\ a_t] \tag{1}$$

Here, the environment is specified as E, the r_t and s_t are the feedback from the environment E and the a_t is the action following the learnt policy π. In short, in reinforcement learning, the reward r_t is used to model the expected return R_t.This expected reward is then placed into the Eq. (1) for the calculation of the Q value. The recursive form of (1) is defined as the bellman Eq. (2) and allows for the estimation of the Q value to be done in similarly to dynamic programming [2].

$$\mathbb{E}_{r_t, s_{t+1} \sim E}\left[r_t + \gamma \mathbb{E}_{a_{t+1} \sim \pi}\left\{Q^{\pi}\left(s_{t,+1}\ a_{t+1}\right)\right\}\right] \tag{2}$$

K. Yada et al. (Eds.): JSAI 2020, AISC 1357, pp. 97–108, 2021.
https://doi.org/10.1007/978-3-030-73113-7_9

There are some notable constraints that come with conventional reinforcement learning. Since the reward is required to calculate the Q value in reinforcement learning, when it is not or cannot be easily defined, reinforcement cannot be conducted. Furthermore, it is also challenging to apply reinforcement learning into practical problems, as we observe firsthand just how subjectively the reward is defined, as the reward is idealized. This amplifies the ambiguity and becomes a pronounced obstacle in complex environments. Researchers have tried to tackle this through the implementation of domain knowledge into reinforcement learning, the so-called inverse reinforcement learning (IRL). By estimating the reward of the actions $Q_D(s, a)$ taken by the demonstrator's domain knowledge, it is possible to eradicate the need to obtain the environment's reward. Incorporating domain knowledge into policy learning also helps shorten the extensive training times required by reinforcement learning due to the high sample density [1]. When given a state and an action, the state action value function is given by the expected return shown in Eq. (1).

While reinforcement learning places an emphasis on the policy, inverse reinforcement learning places a focus on the optimization of a better reward function. Furthermore, the IRL reward is obtained externally from the expert's reward simulation model (domain knowledge). The method above does seem to work well, until we encounter a notorious problem in inverse reinforcement learning, the risk of overlearning, and imitation learning which effectively renders this ineffective. We thus, in this work introduce the implementation of various type of domain knowledges on inverse reinforcement learning. To prevent such constraints and mitigate any discrepancies in the quality of the demonstrator trajectories, we conduct a reward inference process using Bayesian inference. The inference of the reward function was performed between the agent's action and expert demonstrations [2]. Utilizing this, we model the reward function, and derive a better policy, π. There are various algorithms proposed in IRL, including the Maximum Entropy Method (Max-Ent) that utilizes a single linear regression model and the Bayesian Neural Network (BNN) which utilizes a distribution of the rewards. In our Max-Ent, we employ either a Q-Table or a neural network based, model-free reinforcement learning algorithm, Q-Learning. The main objective here is to minimize the differences between the demonstrators reward, and agents reward through Eq. (3), the loss function [2, 3].

$$\nabla \boldsymbol{L}(\theta) = \frac{1}{M} \sum_{i=1}^{M} \emptyset(s_i) - \sum_{i=1}^{M} \boldsymbol{P}(s_i|\theta)\emptyset(s_i) \tag{3}$$

Here the $\emptyset(s_i)$ represents state mapping function. M stands for the length of demonstrator's trajectory. $\boldsymbol{P}(s_i|\theta)$ is the vising frequency calculated by the regression parameter θ. Since we use the visiting frequency of both the agent and the demonstrator, the parameters θ for approximating the reward in the form of a linear regression model are updated to effectively make the two rewards (agent and demonstrator), equivalent [4].

When conducting IRL using BNN, we denote the demonstrator's actions as 'evidence' and utilize that to update the prior on reward functions. This method theoretically offers various advantages over conventional methods, as there is no need to specify an optimal policy for the learning agent. There is no need for the demonstrations to be infallible and thus allows for wide range of demonstrations qualities. The prior model is also relatively flexible, allowing multiple demonstrations, and external information when needed. When comparing BNN to the previously introduced Max-Ent algorithm, where the visiting frequency becomes the reward, the reward in the BNN-IRL is producing by feeding both the demonstrators reward (R_D) and agents reward (R_A) into the neural network. By using BNN embedded in PyTorch, we can produce a distribution of the weights, and subsequently a distribution of the rewards. The two rewards above are used to calculate the gradient of the loss function, or the weights (θ) of the neural network, represented in Eq. (4).

$$\nabla_\theta L = \mathbb{E}_{\tau_m \sim \pi*}[\nabla_\theta R(\tau_m|\theta)] - \mathbb{E}_{\tau \sim \pi_\theta}[\nabla_\theta R(r|\theta)] \quad (4)$$

Here the τ_m is the m-th trajectory of sequence of state-action pairs from the demonstrators by following the demonstrator's policy π^*. On the other hand, the τ represents the trajectory of state-action pairs from the learning agent by following the learning policy π_θ. Consequently, the approximation of the gradient when given demonstrated trajectories T_d is given by Eq. (5).

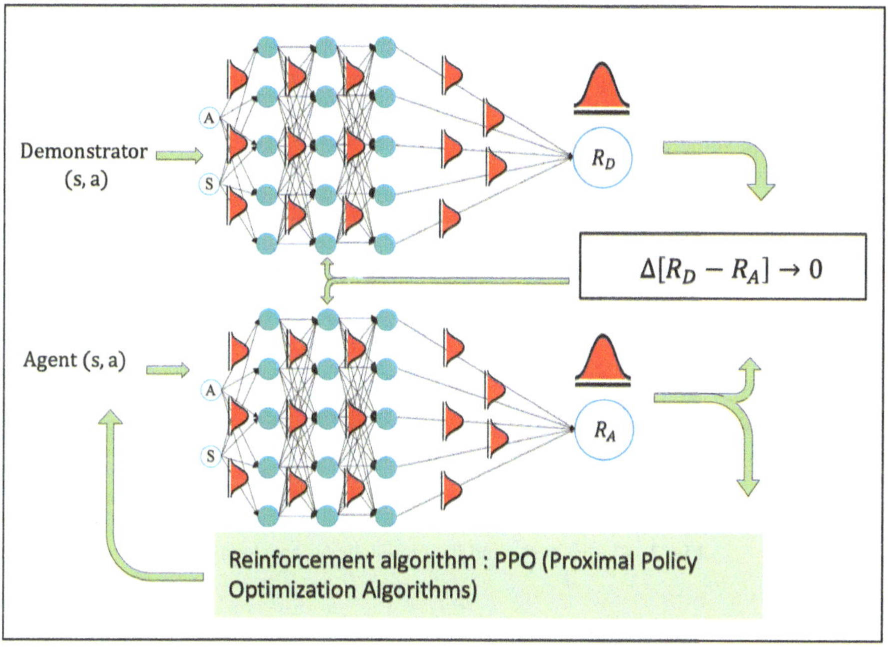

Fig. 1. Schematic of the Bayesian neural network IRL

$$\frac{\partial}{\partial\theta}J(\theta) \approx \frac{1}{|\tau_d|}\sum_{\tau_m \epsilon \tau_d}\sum_{(s,a)\epsilon\tau_m}\frac{\partial}{\partial\theta}r_\theta(s,a) - \frac{1}{|\tau_s|}\sum_{\tau\epsilon\tau_s}\sum_{(s,a)\epsilon\tau}\frac{\partial}{\partial\theta}r_\theta(s,a) + \frac{\partial}{\partial\theta}\log p(\theta) \quad (5)$$

$p(\theta)$ here represents a Gaussian prior distribution over the neural network weights θ. As the reward parameters, and gradients are updated, R_D (the first summation item in Eq. (5) and R_A (the second summation item in Eq. 5) converges to become the same [4]. The agent's new state and action (s, a)' is created by placing the current (s,a) into the reinforcement learning algorithm PPO: proximal policy optimization [5], shown in Fig. 1.

The rest of this paper is organized as follows, section two introduces the experimental setup used to evaluate our primary objective, section three showcases our experimental results through graphs and conceptual descriptions, section four and five state the related works and the conclusion respectively.

2 Experimental Setup

2.1 Mountain Car Environment

The mountain car environment, shown in Fig. 2, is originally a small portion of the cosine curve, with the formula $\cos(3x)$. The movement of the mountain car is controlled by two variables, position x and velcoity v. The y-axis is not required, as it is predetermined, and can be calculated using the equation of the line by substituting into $\cos(3x)$. Instead the velocity of the vehicle is used as the second variable, represented in v. Due to the height of the hill situated on the right, the only solution for the mountain car to successfully reach the goal is initially travel to the hill on the left. After applying full throttle, only then would the vehicle have enough inertia to carry up the slope.

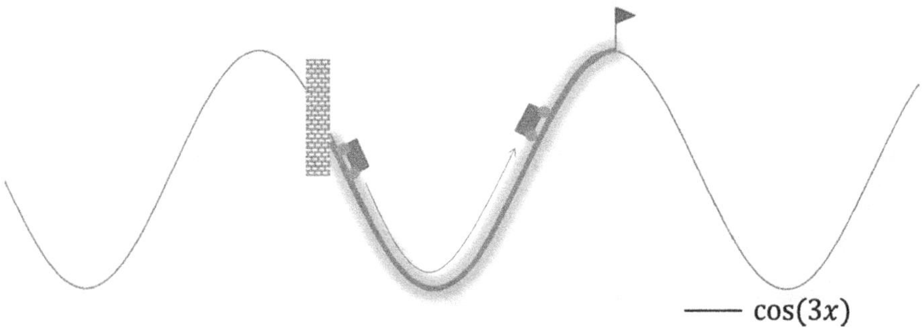

Fig. 2. Cosine curve and the mountain car environment

The possible 'actions' (throttle) that the mountain car can take is defined as follows.

$$Action[-1,\ 0\ ,\ 1]$$

Relative to the x-axis, Action [–1] indicates a throttle direction towards the left, Action [0], indicates no-throttle enforced on the mountain car, and Action [1] is a throttle direction towards the right. Furthermore, the two variables, x (position) and v (velocity) are updated recursively until a specified boundary is reached, shown by the following.

$$Velocity_t = Velocity_{t-1} + (Action) * 0.001 + \cos(3 * Position_{t-1}) * (-0.0025)$$
$$Position_t = Position_{t-1} + Velocity_t$$

Here, the acceleration constant, 0.001 is multiplied to the action, and the gravitational constant −0.0025 is multiplied to the previous position. The recursive update process of x and v is done through the use of t and $t-1$. Thus, when t is the current state, the state before is given by $t-1$. The two boundaries enforced on x and v the following.

$$-1.2 \leq x_{t+1} \leq 0.5$$

When the left bound of position $x = -1.2$ is reached, the velocity resets zero, indicating that the mountain car has hit a wall. When the right bound of the position x is reached, the episode is terminated as the vehicle reached the goal. The boundaries for the velocity of the mountain car, v_t, is written below.

$$-0.07 \leq v_{t+1} \leq 0.07$$

2.2 Demonstrations

In order to determine the impact varying qualities of domain knowledge (demonstrations) would have on the inverse reinforcement learning algorithm, we prepared three different qualities of demonstrators. These include, novice, intermediate and expert demonstrators. The determining factor that differentiates the three qualities is the time (t) taken for each demonstrator reach the goal. When $t \leq 4$ to reach the goal, it is considered as an expert demonstrator. $4 < t \leq 8$ to reach the goal results in the trajectory being categorized as an intermediate demonstrator. Anything beyond $t > 8$ is considered a novice demonstrator. These demonstrations were prepared using the read-char library to read the input from the keyboard. First, we define the keyboard key used as the left arrow key, the right arrow key, and the down arrow key. We also give each key a proper value, since after receiving an input from the keyboard, we can directly forward the action to the environment. The value of the left arrow key is '0', the down arrow key is 1, and the right arrow key is 2. When the user (demonstrator) presses the key, our program first checks if the key pressed is any of the three keys defined above. If so, we use the keyboard's value as action and use this value to get new state from the environment. We collect all the state and action list data for each time step while considering the boundaries of t. The total collected state and action group is counted as one trajectory, and the process

Fig. 3. Schematic of the process of creating demonstration data

repeats until the designated number of the collections of the trajectory is reached. The demonstrations differ in length; thus, we denote the value of state and action as 0 for any unfulfilled length until the limit is reached. The process above is graphically shown in Fig. 3.

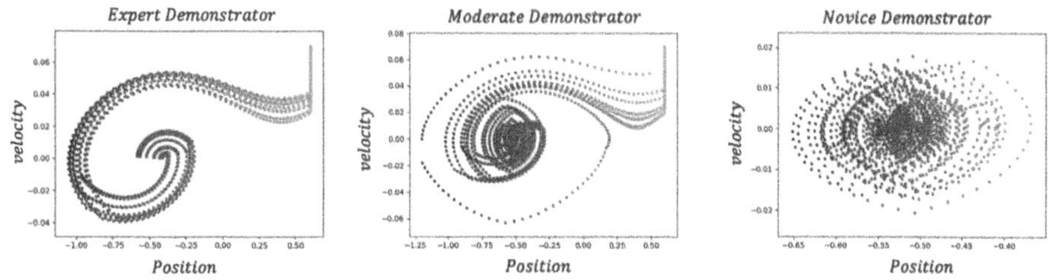

Fig. 4. Novice, intermediate, and expert demonstrators

When 12 of the demonstrations were combined, 3 graphs shown in Fig. 4 is produced. Visually, the differences between each demonstrator evident. We observe a smooth spiral, for the expert demonstrator, indicating that the mountain car efficiently reached the goal, by initially travelling to the left. An inefficient method of reaching the goal is represented in the intermediate demonstrator graph due to the increased concentration of activity towards the center. A clueless behavior is shown with the novice demonstrator, as an elliptical scribble is formed.

2.3 Sim-Reward and Env-Reward

Our research utilized two different rewards, Sim-Reward, and Env-Reward. Sim-Reward is the simulated reward based solely on the behavior trajectory of the demonstrator and is produced by the reward simulation model for the learning process. Env-Reward is the reward obtained from the environment and is the same reward used in conventional reinforcement learning algorithm. It is worthwhile to mention that although conventional reinforcement learning does use the reward from the environment for both learning and evaluation, in IRL, it is only used in the evaluation process, and never in the learning process.

3 Experimental Results

3.1 Comparison Between Bayesian IRL and Max-Entropy IRL

When comparing the scores received when using Bayesian IRL and Max-Ent IRL, we observe from Fig. 5 that the BNN IRL performed significantly better than Max-Ent IRL represented by the higher scores obtained (or towards the right on the x axis). Furthermore, although it is difficult to calculate the average of irregular distributions, we can observe that the maximum probability for bayes is higher than the score at maximum probability for Max Entropy. Furthermore, we observe a comparatively wider distributions of scores for the BNN IRL compared to Max-Ent, most likely because of how BNN uses a distribution of the weights and rewards in the neural network.

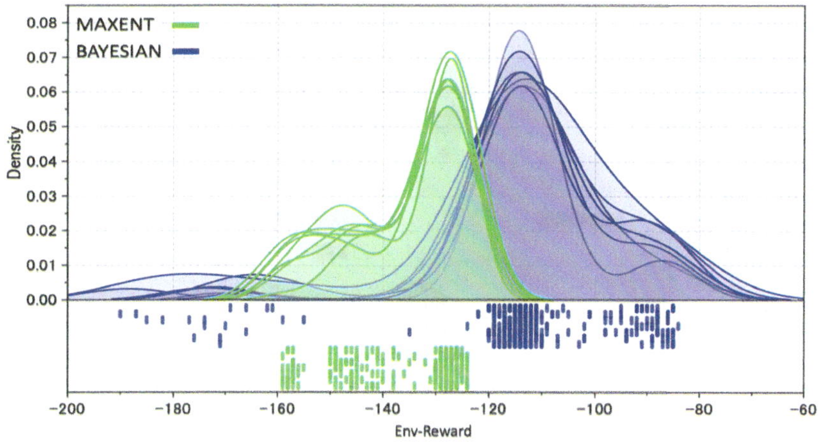

Fig. 5. Comparing Max-Ent and Bayesian IRL

3.2 Cost-To-Go Function

When plotting the value function Q together with the position, x and velocity, y, it produces a three-dimensional model, famously denoted as the cost-to-go function [6]. Here cost is defined as the distance of the mountain car from the goal. In other words, negative of the q value here is the cost. In the initial stages of the learning process, the mountain car is cluelessly moving near in-between the two hills, unable to reach the goal, and thus producing a higher concentration of the q value near the center of the plane. Since the initial value of q is setting as zero for all the state action pairs, thus a small negative value created in the center region drives the agent to conduct more exploration so as to reduce the cost. Therefore observing the resulted moving back and forth. As the learning process nears the end, the agent gradually finds the optimal policy and the number of actions by moving back and forth is reduced and the number of the circles is also reduced correspondingly. This can be seen in the middle learning stage and the end learning stage shown in Fig. 6.

3.3 Simulated Reward Evolution and Distribution During Learning

The algorithm for the reward tuning process was originally designed so that the reward of the learning agent and the demonstrator's reward would become closer. We observed in Fig. 7 that the learning agent initially followed the demonstrator' s actions, which generated similar initial simulated rewards. However, while the episodes are proceeding, the learning agent continued to improve, the learnt reward's profile gradually deviate from the one of demonstrator. This implies there exists an optimal number of learning episodes and in our experiments, the optimal episode is found located around 250.

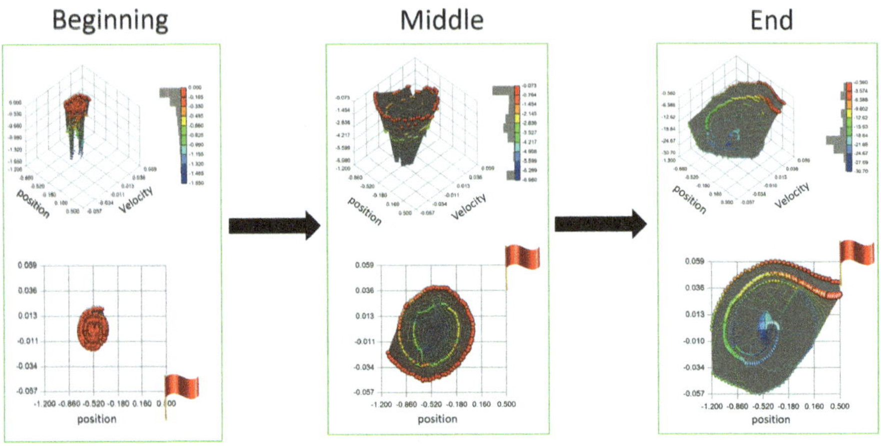

Fig. 6. Comparing Max-Ent and Bayesian IRL.

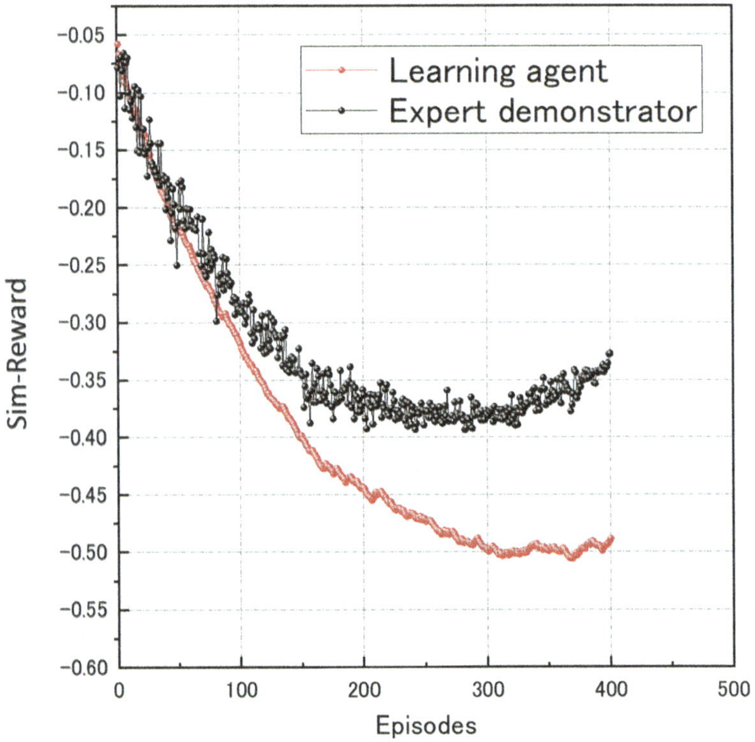

Fig. 7. Comparing the rewards of the learning agent and the demonstrator

For further observation, we recorded the reward distributions of when using two different qualities of demonstrators, an expert demonstrator, and a novice demonstrator in Fig. 8. These were recorded at five different stages of the learning process. Towards

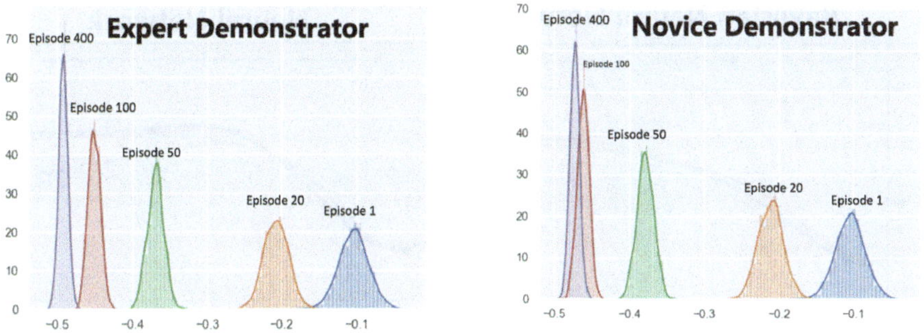

Fig. 8. Reward distributions of Expert(left) and Novice(right) demonstrator

the latter stages of the learning process, we observe that the reward distributions become narrower, as the number of episodes increases. However, we observe no impact the quality of the demonstrator on the distribution of the rewards.

3.4 Comparison Between BNN and Max Entropy Under Different Demonstration Qualities

We observe drastic differences in the scores obtained when using varying qualities of demonstrations on Max-Entropy in Fig. 9. A direct relationship between the quality of the demonstrator trajectories, and the scores obtained from the environment can be seen. The algorithm performed considerably well when using expert demonstrator followed by Intermediate 1, Intermediate 2, and novice demonstrator.

In order to make a more thorough comparison, we converted BNN to a conventional NN by simply removing the variance from the simulated reward distribution. Thus, the

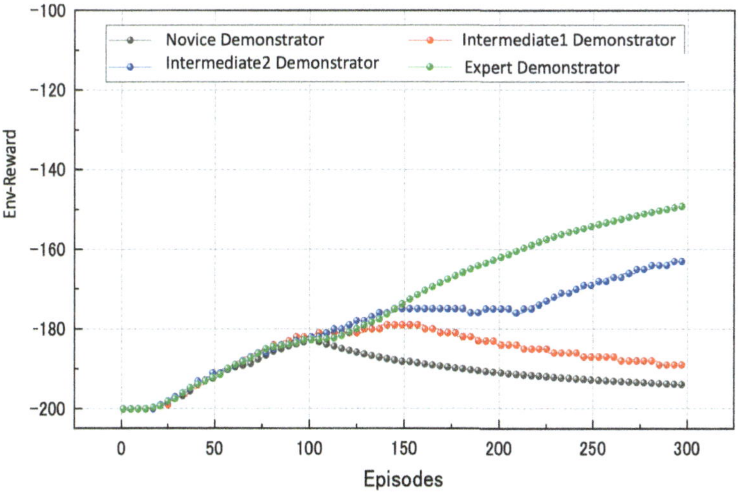

Fig. 9. Demonstrator quality on max entropy

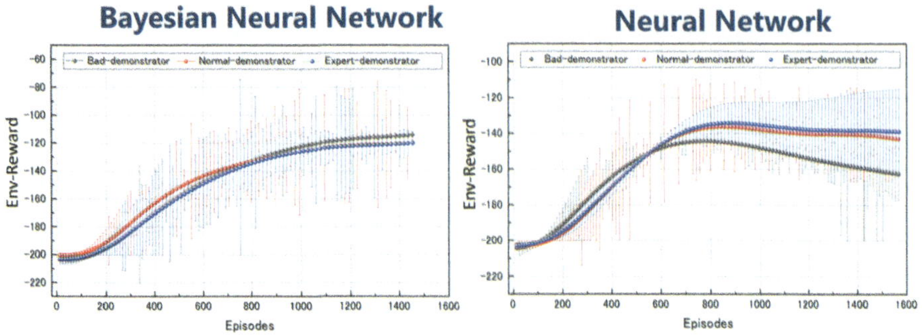

Fig. 10. Comparison between BNN-IRL(left) and NN-IRL(right)

NN rewards were made by essentially taking the mean of the distributions from BNN, which outputs a single reward, other than a reward distribution. With that, when graphing the rewards obtained from the environment of both the Bayesian neural network, and the neural network, in Fig. 10, we observe that the BNN-IRL is capable of achieving similar rewards regardless of the quality of the demonstrator. On the contrary, when using low qualities of demonstrations on the NN, we see a drastic difference in the score obtained.

4 Related Works

Research on inverse reinforcement learning was initially thoroughly done in the latter sections of [7], where IRL was proposed as a method to optimize the reward function, in response to the skepticism surrounding traditional RLs ability in uncertain environments [8] and explored the expression of the reward function as a linear combination of the known ‘features’. Expert demonstrations were used in the IRL algorithm to remove the need to define a specific reward function [9] and discussed the Bayesian approach to implicit imitation in IRL. Bayesian imitation model in [9] allowed the learning agent to smoothly pool prior knowledge, data from the environment, and information referred from the expert's behavior. This differs significantly from traditional imitation reinforcement learning, where the objective is to directly mimic the demonstrator's behavior. On the contrary, Max-Entropy IRL, presenting it as a viable solution to label bias, and noise [10] (to some extent), as it uses a probabilistic approach that maximizes the entropy of the actions. The paper also applies their probabilistic model on real world problems such as modeling routing preferences by drivers to evaluate their proposed method. Acknowledging similar papers noted above, conversely, one of our paper's primary objectives is to comparatively observe the performance of Max-Entropy and Bayesian IRL in an experimental environment such as the Mountain Car with regards to the varying quality of ‘demonstrations’ the agent bases their action on.

5 Conclusion

Due to the difficult nature of implementing domain knowledge into reinforcement learning, we propose inverse reinforcement learning. However, it is difficult to determine to what extent different qualities of demonstrations would impact IRL algorithms' performance (and rewards). Thus, through our research, we found that the differences between Max Entropy IRL and Bayesian IRL can be outlined. Bayesian IRL allows for compromises in the quality of the domain knowledge, as it uses a distribution of rewards and weights. It can thus mitigate uncertainties, as the domain knowledge just is not perfect. This also helps prevent overlearning and imitation from occurring to some extent. On the other hand, we found that max entropy IRL has a general tendency to require high-quality domain knowledge. This is most likely the result of how it uses a single linear regression model, limiting its ability to extract features of the demonstrator. We propose the use of Max Entropy on concise tasks, where there are less factors that the demonstrator have to consider. In other words, Bayesian IRL might prove to play a vital role when placed in a situation where multiple factors would have to be considered for an action to be decided upon, due to its superior ability to extrapolate and interpret the motive behind the demonstrator.

6 Future Scope

As briefly mentioned above, BNN IRL's exceptional representation abilities have shown to be strongly effective compared to other methods in complex environments with multiple factors and multiple target such as optimizing shipping routes for petroleum tankers. Although the single prior distribution for the IRL-BNN algorithm used in this research cannot be applied, by implementing multiple rewards in BNN, thereby increasing the number of prior distributions, such routes scheduling problem can be optimized, even if the environment and situation is considerably more complex. Our team has started working on this project and are optimistic regarding the possibilities Bayesian IRL would have.

References

1. Ramamurthy, R., Bauckhage, C., Sifa, R., Schücker, J., Wrobel, S.: Leveraging domain knowledge for reinforcement learning using MMC architectures. In: Tetko, I.V., Kůrková, V., Karpov, P., Theis, F. (eds.) ICANN 2019. LNCS, vol. 11728, pp. 595–607. Springer, Cham (2019). https://doi.org/10.1007/978-3-030-30484-3_48
2. Oh, M., Iyengar, G.: Sequential Anomaly Detection using Inverse Reinforcement Learning. arXiv:2004.10398 (2020)
3. Wulfmeier, M., Ondruska, P., Posner, I.: Maximum entropy deep inverse reinforcement learning, arXiv:1507.04888 (2015)
4. Nair, A., McGrew, B., Andrychowicz, M., Zaremba, W., Abbeel, P.: Overcoming exploration in reinforcement learning with demonstrations. IEEE Int. Conf. Robot. Autom. (ICRA), 6292–6299(2018)
5. Schulman, J., Wolski, F., Dhariwal, P., Radford, A., Klimov, O.: Proximal Policy Optimization Algorithms. arXiv:1707.06347 (2017)
6. Sutton, R., Barto, A.G.: Reinforcement Learning: An Introduction. MIT Press, MA (1998)
7. Russell, S.: Learning agents for uncertain environments. In: Proceedings of the Eleventh Annual Conference on Computational Learning Theory, ACM, 101– 103 (1998)

8. Abbeel, P., Andrew, N.: Apprenticeship learning via inverse reinforcement learning. In: Proceedings, Twenty-First International Conference on Machine Learning (2004). https://doi.org/10.1007/978-0-387-30164-8_417
9. Price, B., Boutilier, C.: A Bayesian Approach to Imitation in Reinforcement Learning, In: IJCAI 2003: Proceedings of the 18th International Joint Conference on Artificial Intelligence, pp. 712–717 (2003)
10. Ziebart, B.D., Maas, A., Bagnell, J.A., Dey, A.K.: Maximum entropy inverse reinforcement learning. In: Proceedings of the Twenty-Third AAAI Conference on Artificial Intelligence, 1433–1438 (2008)

Intrinsically Motivated Lifelong Exploration in Reinforcement Learning

Nicolas Bougie[1,2(✉)] and Ryutaro Ichise[2,1]

[1] The Graduate University for Advanced Studies, Sokendai, Tokyo, Japan
[2] National Institute of Informatics, Tokyo, Japan
{nicolas-bougie,ichise}@nii.ac.jp

Abstract. Long-term horizon exploration remains a challenging problem in deep reinforcement learning, especially when an environment contains sparse or poorly-defined extrinsic rewards. To tackle this challenge, we propose a reinforcement learning agent to solve hard exploration tasks by leveraging a lifelong exploration bonus. Our method decomposes this bonus into a short-term and a long-term intrinsic reward. The former deals with local exploration - exploring the consequences of short-term decisions, while the latter explicitly encourages deep exploration strategies by remaining large throughout the training process. As formulation of intrinsic novelty, we propose to measure the reconstruction error of an observation given its context to capture flexible exploration behaviors characterized by different time horizons. We demonstrate the effectiveness of our approach in visually rich environments in Minigrid, DMLab, and Atari games. Experimental results validate that our method outperforms baselines in most tasks in terms of scores and exploration efficiency.

Keywords: Reinforcement learning · Exploration · Curiosity in reinforcement learning · Lifelong reinforcement learning

1 Introduction

Reinforcement learning (RL) approaches work by maximizing the excepted rewards provided by the environment. Even though it enables the agent to learn effective policies when the environment has dense rewards, exploration remains a major challenge for environments with sparse rewards or large state spaces. Thus, RL agents often require a large number of interactions to reach decent performance, which can be intractable in real-world settings. A common technique to tackle this problem is to add entropy regularization [17] or to adopt ϵ-greedy exploration [24]. However, such techniques are unlikely to scale with the degree of sparsity of hard exploration tasks.

Recent approaches have proposed to provide intrinsic rewards to guide exploration. This idea is inspired by curious behaviors in animals - visiting novel states should be rewarded. One category of work is count-based exploration [4] in which the agent keeps visit counts for states and encourages the exploration

K. Yada et al. (Eds.): JSAI 2020, AISC 1357, pp. 109–120, 2021.
https://doi.org/10.1007/978-3-030-73113-7_10

of novel states. Another solution [12] relies on information gain to assess novelty. Curiosity can also be measured as the error in predicting the consequences of the agent's actions on the environment. For instance, ICM [19] learns a latent representation of the state and estimates the intrinsic reward based on the error of predicting the next state in the learned latent space. This idea was then extended to capture motion features between consecutive frames [28]. To avoid getting stuck in states with stochastic transitions, several recent works [8,21] have been developed. The intrinsic incentive can also be the combination of a fast and slow intrinsic reward [7]. Despite their ability to improve local exploration - capturing the consequences of short-term decisions, lifelong learning is beyond the reach of most prior work. One reason is the "vanishing curiosity" issue [20,21] inherent in curiosity. It was shown that the novelty of a state tends to vanish quickly with further visitations [8]. As a result, the agent has no long-term incentive to move away from known regions, failing in sparse reward tasks that involve lifelong exploration.

We present an exploration bonus that flexibly combines local exploration and deep exploration (also called long-term exploration). The local exploration bonus encourages the agent to revisit states in the neighborhood of the agent, but swiftly downmodulates states that become more familiar across episodes. On the other hand, deep exploration maintains an incentive across episodes to encourage exploration of hard-to-learn or distant regions - in-depth exploration. The combination of local and deep exploration phases provides an efficient lifelong learning signal to the agent. Local novelty and deep novelty are both estimated based on the reconstruction errors of a *reconstructor* network that takes the current observation's *context* (an observation with missing or corrupted regions) and the executed action, and reconstructs the next frame. The intuition behind this approach is that changing how are created the contexts of the local and deep reward models (i.e. having two types of reconstruction tasks) results in different exploration behaviors that capture environmental dynamics and visual features at different scales (e.g. character-level or word-level). As a side benefit, using noisy inputs forces the model to capture meaningful visual features and is more robust to small changes in the environment.

We benchmark our approach on a set of hard exploration tasks from Minigrid, DMLab, and Atari games. The results show that our technique can escape from local optima in sparse reward environments, to learn comparable or superior policies in most of the tasks.

2 Related Work

The problem of motivating the agent to explore sparse reward environments was studied in the works that we discuss in this section. Most techniques can be grouped into two classes: count-based strategies and curiosity-driven exploration.

The count-based strategy suggests to keep visit counts for states to encourage exploration of rarely visited states [4,15,23]. This idea was then extended to continuous observation spaces. For instance, an approach [18] trains an observation density model to supply counts. A notable step in this direction was taken by a work [1] which keeps counts of clusters rather than states, reducing the size of the count table. It is also possible to discretise the state space via a hash function [25]. Another prior work [16] introduces a count-based optimistic algorithm by estimating the uncertainty associated with each state. Learning without extrinsic rewards has also been studied and is referred to as *novelty search* [14]. The novelty is estimated as the distance of the current behavior features to the *k-nearest* neighbors in a memory of behavior features. Nevertheless, the inherent *uniform exploration* nature of these algorithms might produce inefficient learning when some particular states require more attention. This becomes problematic in tasks that involve lifelong learning since hard-to-reach regions of the state space should be given more weight to ensure enough revisits. On the other hand, our formulation of curiosity automatically scales the exploration bonus based on the complexity (in terms of visual features and dynamics) of the state, and explicitly incentivizes revisits of distant areas.

The proposed method belongs to the category of approaches that consider curiosity as a drive to explore. A number of novelty measurement strategies have been proposed in the past few years, such as ICM [19] that exploits prediction errors of dynamic forward and inverse models. This idea was then extended to include motion features of objects extracted from consecutive frames [28]. A slightly different technique relies on the prediction error in the feature space as measure of the importance of states [13]. In GoCu [6], the authors formulate curiosity as the ability of the agent to achieve a set of goals. By doing so, they incentivize the exploration of hard-to-learn goals. However, it was later shown that the agent can get stuck in local optima due to randomness in the environment [8]. This observation has motivated several methods where the curiosity problem is a deterministic function of its inputs [8,21]. A lifelong learning signal can also be used to improve exploration in Atari games [2]. They simultaneously learn multiple policies with different trade-offs between exploration and exploitation, and encourage lifelong exploration using RND [8]. In this paper, we introduce a different formulation of curiosity that explicitly discards small details to ensure a lifelong learning signal that favors deep exploration. The present work further differs by downmodulating local states that become more familiar to improve exploration efficiency. Finally, our formulation of novelty can be easily tuned to achieve exploration behaviors with different ranges of time horizons. Another method [10] employs a temporal distance predictor and then estimates the novelty. In FaSo [7], the intrinsic reward is the combination of a fast and a slow reward. This work extends this idea; we introduce a different strategy to flexibly combine intrinsic rewards - the deep signal modulates the local novelty reward. This novel strategy alleviates the need for weighting reward streams while effectively promoting lifelong learning. We also aim to develop means for the better incorporation of action-dependent information into existing intrinsic reward formulations. To do so, the new formulation integrates the agent's action and relies on next-frame reconstruction errors. This is critical since large changes

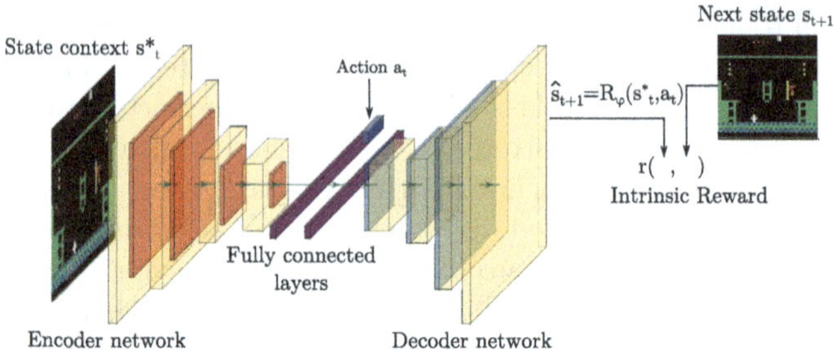

Fig. 1. The reconstructor R_ϕ architecture. The current observation's context and the action are passed through an encoder network and the decoder network attempts to reconstruct the original next frame. Novelty is measured as the discrepancy between the original next frame and the reconstructed next frame.

in the environment can lead to minor visual changes. Finally, we conduct more extensive evaluations of our algorithm on two different domains.

3 Method

In this paper we consider that, along with an extrinsic reward r_t^e, the agent is augmented with an intrinsic reward r_t^i. Consequently, the agent tries to maximize the sum of the intrinsic and extrinsic reward: $r_t = r_t^e + \alpha r_t^i$, where α is a positive scalar weighting the importance of both rewards. In this section, we present the proposed intrinsic reward.

Our main motivation is to develop a lifelong exploration bonus that is able to: 1) capture the importance of states based on the agent's understanding of visual features and motion dynamics, 2) rapidly adapt to assess the novelty of local states, hence leading to fast local exploration, 3) remain large throughout the agent's training to explicitly promote deep exploration behaviors.

To do so, our formulation of curiosity relies on two novelty components. A *local reward* r_t^{local} component that assesses the immediate novelty of each state, rapidly decreasing as the state becomes familiar. On the other hand, a *deep exploration* r_t^{deep} component acts as a deep exploration bonus to explicitly favor deep exploration behaviors. This component remains large across episodes and is maximized by visiting states that induce a significant shift in the state distribution (e.g. discovering a new room), moving the agent toward hard-to-learn regions of the state space. We formulate both novelty bonuses as the error in an agent's ability to reconstruct the next observation based on the current observation's *context* and the agent's action, *context-driven intrinsic curiosity* (Fig. 1).

In the following section, we first describe how is calculated a context-driven intrinsic curiosity component. Next, we formalize how *local* and *deep* exploration components are combined to achieve efficient lifelong exploration.

3.1 Context-Driven Intrinsic Curiosity

We propose a measure of intrinsic motivation formulated as the quality of the agent to reconstruct the next state given the current observation's *context* and the executed action. In this paper, the *context* of an observation refers to a version of it with one or more regions noisy, or corrupted. The pixels can be recovered by propagating local features and capturing the overall structure of the context. The proposed architecture can be understood as a form of forward model, however, a key difference lies in the use of *context* that discards irrelevant details and forces the model to capture meaningful visual features and salient environmental dynamics at different scales (e.g. character-level or word-level), leading to different exploration behaviors characterized by different time horizons. Besides, we found that reconstructing an observation given its context is robust to noisy data or small changes in the environment, helps to capture important visual features, and alleviates the need for modelling hardly predictable fine-grained details. The discrepancy between the reconstructed image and the actual image then serves as the intrinsic reward.

Context Creation. We present a simple method to extract the context of an observation based on the idea of *image downsampling*. The original image of size $w \times w$ is downscaled using a nearest-neighbor interpolation [11] to a smaller image of size $\frac{w}{K} \times \frac{w}{K}$ and then upscaled to the original size, introducing small artifacts. The hyperparameter K controls the amount of artifacts. As an intuition, it can be viewed as a simplified version of the image.

Reward Calculation. At every time step t, the module takes as input the current observation s_t, the next state s_{t+1}, and the action a_t executed at time t. Let s_t be the original observation and s_t^* its context. The reward calculation involves a *reconstructor* network $R_\phi : s^* \times a \mapsto s$ parameterized by ϕ that reconstructs the next state s_{t+1} given the context of the observation and the current action. Write $\hat{s}_{t+1}$ denotes the reconstructed image. Formally, the reconstruction task can be expressed as:

$$\hat{s}_{t+1} = R_\phi(s_t^*, a_t) \tag{1}$$

The reconstruction will have some errors that can be interpreted as a measure of novelty and serves as an intrinsic motivation signal to the agent, r^{cd}. One issue with using reconstruction error as an intrinsic motivation is that the reward function is non-stationary and its scale can fluctuate greatly between different points in time. To keep a consistent scale of the intrinsic rewards, it is useful to normalize the intrinsic reward, therefore, the expression of r^{cd} can be expressed as follows:

$$r^{cd}(s_t, a_t) = \left[\frac{SSIM(s_{t+1}, \hat{s}_{t+1})}{\sigma(R_{cd})}\right] = \left[\frac{SSIM(s_{t+1}, R_\phi(s_t^*, a_t))}{\sigma(R_{cd})}\right] \tag{2}$$

where $SSIM$ is the single-scale SSIM metric [26], $\sigma(R_{cd})$ is the running estimate of the standard deviations of the sum of discounted intrinsic rewards, and s_t^* is

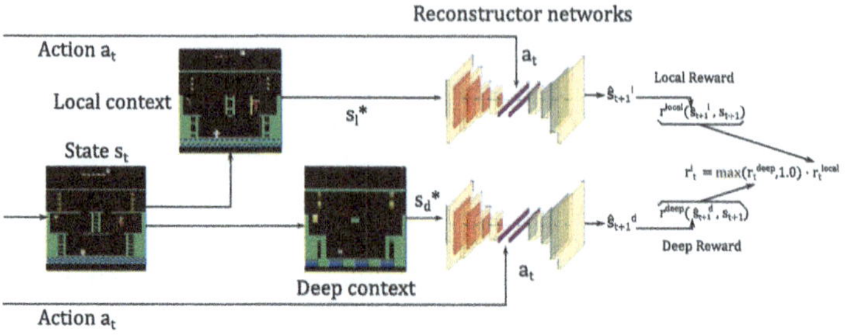

Fig. 2. Training architecture of the proposed model. The reconstructor networks are trained separately based on the local and deep contexts.

the context of the state s_t. Please note that since our method operates in the pixel space, we found that the euclidean distance is ineffective to distinguish noisy data (such as contexts) from the original images. Instead, we propose to embrace the SSIM metric [26], a popular technique used in the field of computer vision for comparing the structural similarity of two images. This formulation is motivated by the idea that, as our ability to model the salient features and environmental dynamics of a particular state improves, the agent has a better understanding and hence the intrinsic motivation is lower.

3.2 Intrinsically Motivated Lifelong Exploration (IML)

In the previous section, we described how is calculated a *context-driven intrinsic curiosity* component. Although a single bonus can improve local or deep exploration, achieving an optimal trade-off between these two behaviors remains a challenging problem [7]. In detail, continuously favoring efficient local exploration will promote the agent to fully explore easy-to-reach states by rapidly adapting the novelty signal. However, the intrinsic reward will exhaust quickly, therefore the agent might struggle to explore hard-to-reach regions. On the other, continuously favoring efficient deep exploration might result in partial exploration of the environment, which entails that the agent can get stuck in local optima and hence converges prematurely to a sub-optimal policy.

Our solution is to decompose the exploration bonus into two components build upon context-driven curiosity (Fig. 2). The *local reward* component computes an intrinsic reward that promotes the visit of surprising states, encouraging local exploration. The *deep exploration* component provides a long-term exploration signal. It is used to scale the reward computed by the local reward component. This scaling does not vanish quickly over time but instead remains large across episodes. The resulting intrinsic reward, r^i_t, efficiently adjusts the trade-off between local and deep exploration phases, facilitating lifelong learning.

The *local reward* component and the *deep exploration* component are calculated by two distinct context-driven curiosity models, which reconstruct the original next state s_{t+1} given the local s^*_l and deep s^*_d contexts of the current state and the executed action, respectively, $R_{\theta_l} : s^*_l \times a \mapsto s$ and $R_{\theta_d} : s^*_d \times a \mapsto s$;

with trainable parameters θ_l and θ_d. An important difference lies in the creation of contexts, s_l^* and s_d^*, to achieve exploration behaviors with different ranges of time horizons. Let K_{local} the parameter used to create the local contexts, we typically used $K_{local} \times 2$ or $K_{local} \times 4$ to create the deep contexts (K_{deep}) - deep contexts are more corrupted. Using more noisy contexts drives the model to discover long-term exploration strategies: 1) as the reconstruction task becomes more challenging when large regions of states are noisy or corrupted, deep reward remains large even with further state visitations, 2) because the reconstructor network will not see fine-grained local differences, it drives the agent to seek out richer and novel regions that induce a significant shift in the state distribution (e.g. visiting a new room), maximizing the deep curiosity signal. On the other hand, contexts of local rewards are nearly unique which entails that slightly deviating from previous policies - visiting novel states or revisiting surprising states (but potentially not fully known), is sufficient to significantly increase the reconstruction errors. Such a local reward decreases quickly across episodes in states that become more familiar in terms of visual features and environmental dynamics.

Thus, the overall intrinsic reward provided by IML is calculated as:

$$r_t^i(s_t, a_t) = max(r_t^{deep}, 1.0) \cdot r_t^{local} \tag{3}$$

This can be re-written as follows:

$$r_t^i(s_t, a_t) = max(r_t^{deep}, 1.0) \cdot \left[\frac{SSIM(s_{t+1}, R_{\theta_l}(s_l^*, a_t))}{\sigma(R_{cd}^l)}\right] \tag{4}$$

where r_t^{deep} is the deep exploration bonus computed by the deep component: $[SSIM(s_{t+1}, R_{\theta_d}(s_d^*, a_t))/\sigma(R_{cd}^d)]$. Intuitively, modulating the local reward by the deep reward results in a frequent switch in regimes of local and deep dominant phases, encouraging the agent to repeatedly revisit not fully explored states in its environment as well as novel regions.

4 Experiments

We test our method in multiple environments from Minigrid [9], DMLab [3], and Atari [5] games. The experiments in Minigrid and DMLab allows us to verify that our method can improve learning in sparse reward environments and its generality. The experiments in Atari games allows us to evaluate our method on tasks that require lifelong exploration.

Environments. In each task, observations are given to the agent in the form of RGB images. We use 84×84 images as input of our method. The agent can choose among a discrete set of actions. We also use a standard frameskip wrapper that repeats each action 4 times (for DMLab and Atari).

RL Algorithm. We selected the commonly used PPO [22] algorithm as our learning algorithm. We employed PPO with similar hyperparameters as in the

Table 1. Final mean score (mean ± std) of our method and baselines on Atari games. We report the results achieved over total 600M timesteps of training, averaged over 10 seeds.

Method	Maximum Mean Score (at convergence)					
	Montezuma's Revenge	Private Eye	Gravitar	Pitfall	Seaquest	Solaris
PPO [22]	2,487 ± 942	103 ± 56	3,438 ± 412	−31 ± 5	1,548 ± 341	1,387 ± 248
PPO+RND [8]	8,152 ± 653	8,666 ± 1051	3,906 ± 246	−3 ± 1	3,179 ± 378	**3,282 ± 281**
PPO+EC [21]	8,025 ± 770	9,244 ± 634	3,521 ± 246	−12 ± 1	**4,650 ± 358**	3,014 ± 273
PPO+ICM [19]	329 ± 118	485 ± 71	3,447 ± 242	−15 ± 2	2,165 ± 223	1,330 ± 108
PPO+FaSo [7]	9,651 ± 442	13,423 ± 775	3,656 ± 280	247 ± 28	4,589 ± 311	3,154 ± 217
Average Human [27]	4,753	69,571	3,351	6,464	20,182	12,327
PPO+IML (ours)	**16,135 ± 688**	**18,004 ± 455**	**6,125 ± 550**	**2,511 ± 373**	3,814 ± 504	3,120 ± 338

original implementation of RND [8]. The output of the last convolutional layer is fed into a fully connected layer with 256 units. It is followed by two separate fully connected layers of size 512, used to predict the value function of each reward component (extrinsic and intrinsic advantage).

Baseline Methods. The simplest baseline is PPO trained without intrinsic reward. We also compare with the following baselines: ICM [19], RND [8], FaSo [7], and EC [21]. We used same hyperparameters as in the original implementations.

Implementation Details. On DMLab, we employ contexts created with $K_{local} = 2$ and $K_{deep} = 5$. On Atari and Minigrid, we compare our method trained with contexts $K_{local} = 2$ and $K_{deep} = 4$. We set the coefficient of intrinsic reward $\alpha = 0.5$. The architecture of the reconstructor networks consists of a sequence of four convolutional layers with 32 filters each, stride: 2,1,1, kernel size of 3×3, and padding 1. We apply a rectifier non-linearity after each convolutional layer. For online training, we store the experience and make 5 epochs of training every 12K steps. We found that retraining the *deep* reconstructor network once every 30K steps is sufficient. Note that we decouple the training speed of the reconstructor models to ensure that the models achieve the desired behaviors. Training is carried out with a fixed learning rate of 0.0003 using the Adam optimizer.

4.1 Hard Exploration Games

We first conduct experiments on six established hard exploration Atari 2600 games from the Arcade Learning Environment (ALE) [5] that feature very sparse rewards and complex visual features. Moreover, solving these tasks requires a long-term exploration bonus to escape from sub-optimal policies by compensating the loss of easy immediate extrinsic reward and by incentivizing useful decisions on a long-time horizon. For instance, on Montezuma's Revenge, to solve the first level the agent must give up easy to reach immediate rewards associated with "opening a door" and save the key useful later to open a door. Therefore, curiosity should remain large enough to compensate the loss of immediate extrinsic reward and thus promote the discovery of long-term strategies (potentially leading to a higher return).

Table 2. Scores (mean ± std) on tasks from the DMLab domain and Atari benchmark with dense and sparse settings. "Task 1" denotes "Rooms Collect Good Objects Train" and "Task 2" denotes "Room Keys Doors Puzzle". The results are averaged over 15 runs after 1200 steps (DMLab) or 600M steps (Atari).

Method	Sparse		Dense		Atari	
	Task 1	Task 2	Task 1	Task 2	Pong	Breakout
PPO	9.34 ± 0.03	25.0 ± 0.52	9.61 ± 0.02	26.8 ± 4.20	9.02 ± 1.86	10.5 ± 2.38
PPO+RND	9.96 ± 0.05	30.15 ± 2.26	9.86 ± 0.03	28.8 ± 0.70	10.3 ± 0.08	424 ± 21.4
PPO+EC	9.68 ± 0.03	**33.2 ± 0.56**	9.61 ± 0.03	**32.8 ± 0.54**	9.19 ± 1.27	31.1 ± 6.94
PPO+ICM	9.55 ± 0.02	21.8 ± 0.41	9.52 ± 0.02	20.5 ± 0.69	8.11 ± 0.06	438 ± 5.88
PPO+Faso	9.59 ± 0.04	23.4 ± 0.78	9.56 ± 0.04	22.7 ± 0.66	10.5 ± 0.07	416 ± 6.32
PPO+IML (ours)	**9.98 ± 0.03**	29.9 ± 0.65	**9.92 ± 0.02**	28.7 ± 0.59	**14.6 ± 1.12**	**620 ± 12.5**

The average scores are reported and compared in Table 1. It can be observed that IML outperforms the baselines in Montezuma's Revenge, Gravitar, and Pitfall, and Private Eye. In Seaquest, the scores of our method is comparable to those of PPO+Faso and PPO+EC. The rationale is that these games demand long-term exploration incentives, and thus are harder to be solved by using a curiosity-based bonus that vanishes quickly. For instance, on Pitfall the agent must go through regions of the state space that can yield negative rewards, which dissuade baselines from exploring efficiently. On the other hand, since our formulation of curiosity does not vanish with further state visitations, it allows our agent to discover alternative ways to obtain the optimal reward.

4.2 Dense Reward Tasks

In this experiment, we aim to investigate the impact of dense rewards on the performance of the proposed method. It is desirable to avoid deteriorate performance in dense reward tasks. We test this idea in two levels in the DMLab environment: "Rooms Collect Good Objects Train" and "Room Keys Doors Puzzle", and two dense reward Atari games: Pong and Breakout. In the first DMLab task, the agent has to pick objects randomly positioned across the map and avoid dangerous objects. In the second DMLab task, the agent has to collect keys useful to open a serie of doors. In all scenarios, the task rewards are quite dense since the agent receives a positive or negative reward after collecting an object or taking a meaningful action. In Table 2, we can see that our method does not significantly decrease the performance. In task 1 and task 2, although IML performs moderately worse in the dense setting, it still drastically improves the score as compared to plain PPO.

4.3 Procedurally Generated Environments

We now validate IML capability of solving procedurally generated environments and generalizing to unseen views or appearances. Figure 3 shows the training curves on Minigrid Door & key of our method against four baselines. We can

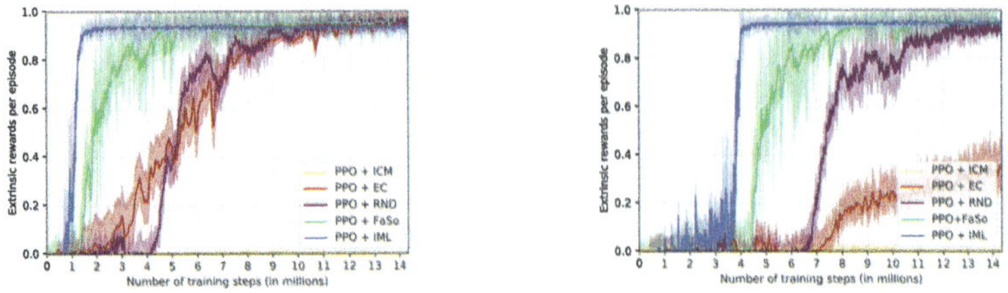

(a) Door & key 11×11 (b) Door & key 16×16

Fig. 3. Average results over 10 random seeds on Minigrid (Door & key). The shaded area shows the standard errors of 10 runs.

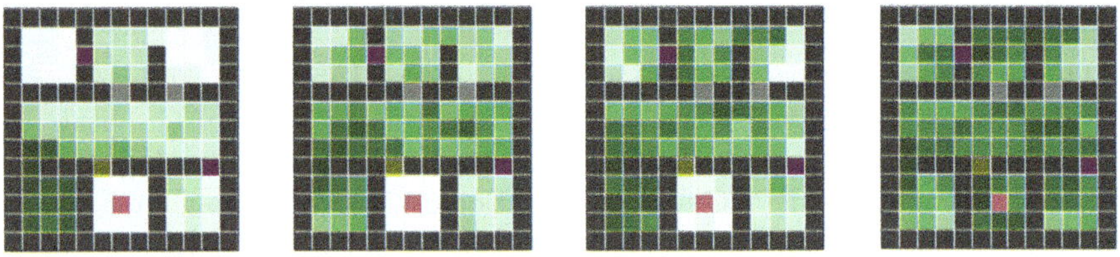

(a) PPO + ICM (b) PPO + RND (c) PPO + FaSo (d) PPO + IML

Fig. 4. State visitation heatmaps averaged over 10 runs for different models: ICM, RND, FaSo, and IML. We trained the models for 10M frames on randomly generated mazes in KeyCorridorS5R3.

observe that PPO tends to perform poorly due to the sparsity of the tasks. From Fig. 3, it is clear that ICM is significantly outperformed, we suspect that the high similarity among states results in small rewards. Moreover, learning effective controllable states using a simple inverse dynamics model can be challenging. On the other hand, RND can learn near-optimal policies but is slower than the proposed method. Overall, the results show that our method can learn effective policies in procedurally generated environments. It confirms that IML is a crucial element that makes learning from sparse rewards possible.

4.4 No Reward/Area Coverage

This experiment aims to evaluate the good exploration coverage of IML when no extrinsic reward is provided. We use KeyCorridorS5R3 from the Minigrid environment and eliminate the reward. The agents only receive a curiosity-based signal. As can be observed in Fig. 4, to remain curious the agent is pushed to explore distant regions of the state space, which entails that it should discover new rooms. The experimental results highlight that in the absence of task rewards, the proposed method provides enough indirect supervision for learning meaningful behaviors. Moreover, we can observe that only IML can fully explore all the rooms.

5 Conclusion

This paper introduced an exploration method for lifelong exploration in sparse reward environments. Our formulation of curiosity flexibly combines a local exploration bonus and a deep exploration bonus that promotes in-depth exploration across episodes. They use the idea of context-driven reconstruction to capture salient visual features and motion dynamics at different scales of the environment. This allows us to overcome the known "vanishing curiosity" issues of prior work by providing an intrinsic reward that remains large during the agent's training. We demonstrated the proposed methodology and compared it against a number of baselines on Minigrid, DMLab, and Atari tasks. Experimental results showed that IML is capable of achieving lifelong exploration, including in tasks with deceptive rewards.

References

1. Abel, D., Agarwal, A., Diaz, F., Krishnamurthy, A., Schapire, R.E.: Exploratory gradient boosting for reinforcement learning in complex domains. In: ICML Workshop on Abstraction in Reinforcement Learning (2016)
2. Badia, A.P., et al.: Never give up: learning directed exploration strategies. In: International Conference on Learning Representations (2020)
3. Beattie, C., et al.: DeepMind lab. arXiv preprint arXiv:1612.03801 (2016)
4. Bellemare, M., Srinivasan, S., Ostrovski, G., Schaul, T., Saxton, D., Munos, R.: Unifying count-based exploration and intrinsic motivation. In: Advances in Neural Information Processing Systems, pp. 1471–1479 (2016)
5. Bellemare, M.G., Naddaf, Y., Veness, J., Bowling, M.: The arcade learning environment: an evaluation platform for general agents. J. Artif. Intell. Res. **47**, 253–279 (2013)
6. Bougie, N., Ichise, R.: Skill-based curiosity for intrinsically motivated reinforcement learning. Mach. Learn. **109**, 493–512 (2019)
7. Bougie, N., Ichise, R.: Combining local and global exploration via intrinsic rewards. In: Annual Conference of JSAI, pp. 2K6ES205–2K6ES205 (2020). https://doi.org/10.11517/pjsai.JSAI2020.0_2K6ES205
8. Burda, Y., Edwards, H., Storkey, A., Klimov, O.: Exploration by random network distillation. In: Conference on Learning Representations (2019)
9. Chevalier-Boisvert, M., Willems, L., Pal, S.: Minimalistic gridworld environment for OpenAI gym (2018). https://github.com/maximecb/gym-minigrid
10. Fu, J., Co-Reyes, J., Levine, S.: Ex2: exploration with exemplar models for deep reinforcement learning. In: Advances in Neural Information Processing Systems, pp. 2577–2587 (2017)
11. Han, D.: Comparison of commonly used image interpolation methods. In: Conference on Computer Science and Electronics Engineering (2013)
12. Houthooft, R., et al.: VIME: variational information maximizing exploration. In: Advances in Neural Information Processing Systems, pp. 1109–1117 (2016)
13. Klyubin, A.S., Polani, D., Nehaniv, C.L.: Empowerment: a universal agent-centric measure of control. In: IEEE Congress on Evolutionary Computation, vol. 1, pp. 128–135 (2005)

14. Lehman, J., Stanley, K.O.: Abandoning objectives: evolution through the search for novelty alone. Evol. Comput. **19**, 189–223 (2011)
15. Machado, M.C., Bellemare, M.G., Bowling, M.: Count-based exploration with the successor representation. In: AAAI Conference on Artificial Intelligence, pp. 5125–5133 (2020)
16. Martin, J., Sasikumar, S.N., Everitt, T., Hutter, M.: Count-based exploration in feature space for reinforcement learning. In: International Joint Conference on Artificial Intelligence (2017)
17. Mnih, V., et al.: Asynchronous methods for deep reinforcement learning. In: Conference on Machine Learning, pp. 1928–1937 (2016)
18. Ostrovski, G., Bellemare, M.G., van den Oord, A., Munos, R.: Count-based exploration with neural density models. In: International Conference on Machine Learning, pp. 2721–2730 (2017)
19. Pathak, D., Agrawal, P., Efros, A.A., Darrell, T.: Curiosity-driven exploration by self-supervised prediction. In: International Conference on International Conference on Machine Learning (2017)
20. Raileanu, R., Rocktäschel, T.: Ride: rewarding impact-driven exploration for procedurally-generated environments. In: International Conference on Learning Representations (2020)
21. Savinov, N., et al.: Episodic curiosity through reachability. In: International Conference on Learning Representations (2019)
22. Schulman, J., Wolski, F., Dhariwal, P., Radford, A., Klimov, O.: Proximal policy optimization algorithms. arXiv preprint arXiv:1707.06347 (2017)
23. Strehl, A.L., Littman, M.L.: An analysis of model-based interval estimation for Markov decision processes. J. Comput. Syst. Sci. **74**(8), 1309–1331 (2008)
24. Sutton, R.S., Barto, A.G.: Reinforcement Learning: An Introduction. MIT Press, Cambridge (1998)
25. Tang, H., et al.: # exploration: a study of count-based exploration for deep reinforcement learning. In: Advances in Neural Information Processing Systems, pp. 2753–2762 (2017)
26. Wang, Z., Bovik, A.C., Sheikh, H.R., Simoncelli, E.P., et al.: Image quality assessment: from error visibility to structural similarity. IEEE Trans. Image Process. **13**(4), 600–612 (2004)
27. Wang, Z., Schaul, T., Hessel, M., Van Hasselt, H., Lanctot, M., De Freitas, N.: Dueling network architectures for deep reinforcement learning. In: International Conference on Machine Learning (2016)
28. Yang, H.K., Chiang, P.H., Hong, M.F., Lee, C.Y.: Flow-based intrinsic curiosity module. In: Proceedings of the the International Conference on Learning Representations (2020)

A Preliminary Analysis of Offensive Language Detection Transferability from Social Media to Video Live Streaming Platforms

Zhiwei Gao(✉), Shuntaro Yada, Shoko Wakamiya, and Eiji Aramaki

Nara Institute of Science and Technology, Ikoma, Japan
{gao.zhiwei.fw1,s-yada,wakamiya,aramaki}@is.naist.jp

Abstract. Given the growing popularity of online games and eSports, the young generation is increasingly enjoying its video live streaming service. Offensive conversations often appear against the streamer or audience in the streaming channel's chatroom. This research aims to detect offensive language appearing in video live streaming chats. Focusing on Twitch, the most popular live streaming platform, we created a dataset for the task of detecting offensive language. We collected chat posts across four popular game titles with genre diversity (i.e., competitive, violent, peaceful). To make use of the similarity in offensive languages among social media, we adopt the state-of-the-art models trained over the offensive language on Twitter to our Twitch data (i.e., transfer learning). Our results show that transfer from social media to live streaming is effective. However, the similarity measures we proposed show less correlation on the transferability prediction.

Keywords: Machine learning · Transfer learning · Social media · Twitch

1 Introduction

Posting offensive comments on social media is a common problem of abusive behavior. An offensive comment is characterized by profanes, insults, curses, and obscenities against one person or a group of people (Alakrot et al. 2018). Unlike a direct talk, in electronic text communication, there is no eye-to-eye contact or sense of personal presence that dictates a certain social etiquette. People who are separated from each other geographically can attack each other verbally without fear of physical harm (Chapman 1995). Due to the negative psychological and physical consequences that may affect adolescents, it is essential to detect offensive language in social media accurately. On the task of detecting offensive language, one of the most popular social media platforms, Twitter, has already attracted a lot of attention due to its large user base and the linguistic idiosyncrasies, such as tendency to use "@" to tag a user or Internet slang and neologism (Matsumoto et al. 2019).

K. Yada et al. (Eds.): JSAI 2020, AISC 1357, pp. 121–132, 2021.
https://doi.org/10.1007/978-3-030-73113-7_11

Table 1. Offensive post examples from Twitter and Twitch. Capitalized words are offensive expressions.

Twitter	@user please don't make the same mistake with #Kavanaugh! He is a **PERJURER** who is also a Trump lackey. #VoteNoOnKavanaugh
	@user Y'all some **BITCHES** lol
	@user let me **NIGGA** fish in peace damn
Twitch	@user I know but she is a great women.,Can't wait to finally **BE YOUR DAD**
	You are such a **FUCKIN** dog
	LMAO THEY ARE SO **STUPID**

It is also important to focus on other growing/upcoming platforms that share similar features with Twitter so as to provide more protection to users, more inspection methods for service providers, and more diversified characteristics for research on offensive language detection.

Twitch[1] is one of the most popular video live-streaming platforms that share similar features with Twitter because it had 2.2 million broadcasters monthly and 15 million daily active users by 2018[2]. Each live channel of Twitch has a chat room for viewers to interact in real-time. Different from Twitter, Twitch's chat cannot be deleted or modified by the sender. Table 1 shows some offensive language examples from Twitter and Twitch, respectively.

Although modern neural architectures for NLP tasks including offensive language detection are highly effective when provided with a large amount of labeled training data (Zhang et al. 2015; Conneau et al. 2017; Bowman et al. 2015), especially for emerging platforms, it is difficult to access large amounts of labeled data. In the context, a special scenario of transfer learning, domain adaptation, aims to learn transferable representation from a source domain and apply it to a target domain. The question in domain adaptation is usually framed as "How to transfer features from a source domain to a target domain?" Based on the domain adaptation, the question we address is "How to use existing datasets to construct a better dataset for transfer learning with limited data?" We propose to make use of (1) an existing large labeled dataset (Twitter dataset) and (2) combinations of large dataset and small datasets with the same size (within Twitch datasets) as the source domain, in order to demonstrate the effectiveness of transferring its features to the target domain. We collected 2,000 chat posts across four popular game titles with genre diversity to construct the small datasets. Furthermore, to compare the impact of different small datasets on transferability, we assumed that the more similar the small datasets are, the higher the performance of the transferability. We considered the similarity of the chat posts and the similarity of target content, which is discussed in Sect. 4.2.

[1] https://www.twitch.tv.
[2] https://en.wikipedia.org/wiki/Twitch_(service).

2 Related Work

For the offensive language detection task, Cheng et al. (2017) found that both negative mood and seeing troll posts by others significantly increases the probability of a user trolling, and together double this probability through an experiment simulating an online discussion. Rezvan et al. (2020) presented experiments with Twitter corpus to analyze harassing language for five types: (i) sexual, (ii) racial, (iii) intellectual, (iv) appearance-related and (v) political. Waseem et al. (2017) presented a typology that synthesizes the different sub-tasks in abusive language detection. The author creatively divides offensive language into explicit and implicit by the way it is expressed, and also into direct and generalized by the objects of offensive language. SemEval-2019 Task 6: Identifying and Categorizing Offensive Language in Social Media (Zampieri et al. 2019b) organized a shared task on offensive language detection on Twitter. The task consisted of three sub-tasks: (a) offensive language identification, (b) categorization of offense types and (c) offense target identification. Their work provides us with a well-annotated dataset. Liu et al. (2019a) and Nikolov and Radivchev (2019) proposed their best performance models for each sub-task.

This study inspired by a previous study, Dai et al. (2019), which shows that the effectiveness of pre-trained language model depends on the source data: if it has a high vocabulary intersection with the target data, the model will be more effective. Guo et al. (2020) proposed that when labeled data from target domain is available, supervised domain adaptation can achieve state-of-the-art results via fine-tuning, especially when source domain has orders of magnitude more data than target domain.

3 Datasets

3.1 Data Collection

Since Twitter and Twitch are very similar in spelling, to avoid confusion, in the following sections we use *social media dataset* to refer to the Twitter dataset, and *live chat dataset* to refer to the Twitch dataset.

Social Media Dataset. To verify the transferability, we used an existing dataset as the *social media dataset*: Offensive Language Identification Dataset (OLID) (Zampieri et al. 2019a). The OLID dataset was used in SemEval-2019 Task 6. The organizers collected tweets including specific keywords and annotated them hierarchically regarding the offensive language, offense type, and offense target. This task consists of three sub tasks. Sub-task A involves detection of offensive posts and is the most relevant to our research. The keywords used include words that point to specifics, such as “you are” and “she is”, as well as politics-related keywords such as “MAGA” and “gun control”. For this sub-task, there are 13,240 tweets in the training data and 860 tweets in the test data.

Table 2. Four different games and their keywords

Game	Description	Keywords
Fortnite (LC_{FN})	Popular survival shooter game	Action, shooter, TPP, tactical, zombie, sandbox, survival, multiplayers, cooperation, crafting
Grand Theft Auto V (LC_{GT})	Open-world game known for its violent content	Action, FPP, TPP, vehicles, gangster, sandbox, shooter, modern, adventure, open-world
Hearthstone (LC_{HS})	Digital collectible card game	Logic, fantasy, card game, 2 players, strategy, miscellaneous, turn-based
Minecraft (LC_{MC})	Sandbox, Lego-like game	Adventure, FPP, TPP, fantasy, sandbox, RPG, elements, survival, multiplayers, cooperation, crafting, action, 3D

Table 3. Four live chat datasets and samples

Dataset	Samples
LC_{FN}	@user YOU ARE SO GOOD IN TARKOV WHY do not YOU PLAY AGAIN
	Wtf are these people you are coming against
	Now he is saying he could 2v1 him
LC_{GT}	YOU ARE BEING HUNTED
	he is to tired :KEKW:
	@user you are so hot even my zipper is falling for you
LC_{HS}	@user it means you just lost your life :kappa:
	If you think he should sell bolvar you are :Pepega:
	He is trading because all face hunters are sttupid
LC_{MC}	He is in the doghouse for all the bits and stuff :kappa:
	so if you are BRONZE GOD does that make Ettie the Sea Goddess
	@user you were supposed to try ash blocks

Live Chat Dataset. For the *live chat dataset*, we expected that the corresponding chats for the different game genres would contain different types of offensive content. Therefore, we chose four different genres of games and then collected data from these games' chat rooms in January 2020. To characterize each game, we assigned "game keywords" obtained from game review sites Metacritic[3] and GamePressure[4]. The game name and keywords are shown in Table 2 and samples of each dataset shown in Table 3.

We tried to use the same keywords as the OLID to collect data. Considering there are far fewer politics-related messages in Twitch than in Twitter, we only

[3] https://www.metacritic.com/.
[4] https://www.gamepressure.com/.

Table 4. Class distribution of each game in *live chat dataset*

	No. of offensive (%)	No. of not Offensive (%)
LC_{FN}	77 (15.4)	423 (84.6)
LC_{GT}	62 (12.4)	438 (87.6)
LC_{HS}	68 (13.6)	432 (86.4)
LC_{MC}	36 (7.2)	464 (92.8)

kept "she is," "he is," and "they are" as the keywords. Since we found out that many users use abbreviations, we also use acronyms such as "u r" (you are) as keywords.

Finally, we collected 2,000 posts from the four game titles and each post was then annotated as offensive or not offensive by two coders. We employed Cohen's kappa (Cohen 1960) to measure inter-coder agreement and achieved 0.77 of Cohen's kappa, which shows a substantial agreement between our coders. Table 4 shows the class distribution of each game.

3.2 Preprocessing

To reduce noise, we converted all texts into lower case and replaced all URL addresses with "http." We used an online emoji project on GitHub[5], which maps an emoji to a phrase (Liu et al. 2019a), so that we could retain as much semantic information as possible without destroying the quality of the sentence. We did this preprocessing on both *social media dataset* and *live chat dataset*.

The live chat dataset also contains a large number of user-created emojis called "Emotes" such as "kappa" and "PepeHands." Since these "Emotes" contain emotion and information, we applied word segmentation using Ekphrasis (Baziotis et al. 2017), a text processing tool built specially for social media posts. An emote code is split into the separate word tokens, provided that there is an upper-cased word within the emote code. For instance, the token "PepeHands" is split into two tokens: "pepe" and "hands."

4 Experiments and Results

4.1 Model

We adopted the pre-trained model BERT (Bidirectional Encoder Representation from Transformer) (Devlin et al. 2019) because it has achieved state-of-the-art results on many NLP tasks, including SemEval-2019 Task 6 on offensive language identification (Liu et al. 2019a). We also adopted the following models: RoBERTa (Liu et al. 2019b), XLNet (Yang et al. 2019), fastText (Joulin et al. 2017), and combined the best of them in ensembles.

[5] https://github.com/carpedm20/emoji.

Since that there are many versions of the aforementioned models, We used the base and uncased versions, i.e., *bert-base-uncased* and *roberta-base*, respectively. We used the Adam optimizer (Kingma and Ba 2014) to optimize the loss function. We trained each model with 5 epochs and set batch size to 64.

Table 5 shows our results of the aforementioned models trained and tested on the *social media dataset*. We selected RoBERTa as our selected model since it achieved the best performance. Named RoBERTa for "Robustly Optimized BERT approach," the model adopts many of the techniques used by BERT. As it is developed based on BERT, they share lots of configurations. The following items are different between RoBERTa and BERT (Liu et al. 2019; Devlin et al. 2019):

- **Reserved Token:** BERT uses [CLS] and [SEP] as starting token and separator token respectively while RoBERTa uses < s> and </s> to covert sentences.
- **Size of Subword:** BERT has around 30k subwords while RoBERTa has around 50k subwords.

Table 5. Results on social media test dataset.

	Precision	**Recall**	**F1-Score**
BERT	0.793	0.781	0.787
RoBERTa	0.792	0.798	**0.795**
XLNet	0.782	0.785	0.784
FastText	0.684	0.687	0.686

4.2 Similarity Measures

Similarity of Chat Posts: Since similar chat posts contribute to transfer learning, we employed Target Vocabulary Covered (TVcC) to measure the similarity of our *live chat dataset*. TVcC shows the percentage of the target vocabulary that is also presented in the source datasets and only calculate the content words (nouns, verbs, adjectives) (Dai et al. 2019). TVcC is the most informative measure in predicting the performance of pre-trained language models given a target dataset among four measures proposed by Dai et al. (2019). It is calculated as follows:

$$TVcC\left(D_{\mathrm{S}}, D_{\mathrm{T}}\right) = \frac{\left|V_{\mathrm{D_S}} \cap V_{\mathrm{D_T}}\right|}{\left|V_{\mathrm{D_T}}\right|},$$

where V_{D_S} and V_{D_T} are sets of unique tokens in source an target datasets respectively.

Table 6. Results of RoBERTa model using *live chat datasets*. The left column shows the test set. In the training set columns, the training set used in the current test set is checked. We used "*Twitter + X*" to denote our training sets, where Twitter is the baseline, and "*X*" refers to its combinations with LC_{FN}, LC_{GT}, LC_{HS}, and LC_{MC}. The TVcC column is the similarity of chat posts. The overlap coefficient column is the similarity of targets (games).

Test set	Training set: Twitter	$+LC_{\text{FN}}$	$+LC_{\text{GT}}$	$+LC_{\text{HS}}$	$+LC_{\text{MC}}$	TVcC (%) (without Twitter)	Overlap coefficient (%)	Precision	Recall	F1-score
Twitter	✓					-	-	0.792	0.798	0.795
LC_{FN}	✓					-	-	0.766	0.832	0.792
	✓		✓			36.30	40	0.784	0.802	0.793
	✓			✓		32.72	0	0.773	0.808	0.788
	✓				✓	34.89	70	0.805	0.791	0.798
	✓		✓	✓		47.21	40	0.782	0.794	0.788
	✓		✓		✓	47.46	80	0.822	0.787	**0.802**
	✓			✓	✓	45.55	70	0.800	0.786	0.792
	✓		✓	✓	✓	**53.71**	80	0.806	0.774	0.788
LC_{GT}	✓					-	-	0.686	0.780	0.715
	✓	✓				34.94	40	0.712	0.781	0.738
	✓			✓		30.61	0	0.717	0.791	0.745
	✓				✓	32.13	50	0.719	0.752	0.734
	✓	✓		✓		44.55	40	0.722	0.780	0.745
	✓	✓			✓	44.23	60	0.734	0.745	0.739
	✓			✓	✓	42.55	50	0.751	0.769	**0.759**
	✓	✓		✓	✓	**50.40**	60	0.736	0.761	0.747
LC_{HS}	✓					-	-	0.729	0.778	0.749
	✓	✓				31.54	0	0.749	0.768	0.758
	✓		✓			30.66	0	0.764	0.770	**0.768**
	✓				✓	30.50	14	0.778	0.749	0.762
	✓	✓	✓			41.17	0	0.773	0.763	0.767
	✓	✓			✓	40.77	14	0.794	0.742	0.764
	✓		✓		✓	40.93	14	0.790	0.740	0.761
	✓	✓	✓		✓	**46.95**	14	0.789	0.741	0.762
LC_{MC}	✓					-	-	0.611	0.777	0.632
	✓	✓				33.47	70	0.620	0.765	0.648
	✓		✓			32.03	50	0.630	0.767	0.660
	✓			✓		30.35	14	0.627	0.772	0.657
	✓	✓	✓			42.73	69	0.632	0.771	0.664
	✓	✓		✓		42.65	62	0.630	0.763	0.660
	✓		✓	✓		42.25	46	0.633	0.756	0.664
	✓	✓	✓	✓		**48.48**	77	0.639	0.770	**0.672**

Similarity of Target Contents: Considering a similar target (game) content would have an impact on offensive posts. For instance, violent or competitive games perhaps lead to more offensive behaviors. Because keywords of a target are a concise overview of its content, we employed an overlap coefficient of the target review keywords shown in Sect. 3 to get the similarity between target contents. It is calculated as follows:

$$overlap\,(R_{\text{A}}, R_{\text{B}}) = \frac{|R_{\text{A}} \cap R_{\text{B}}|}{min\,(|R_{\text{A}}|\,, |R_{\text{B}}|)},$$

where R_A and R_B are the keywords for two different targets (games). Higher overlap coefficient indicates higher similarity of the two targets.

4.3 Evaluation and Results

We evaluated the performance of our models based on precision, recall and macro F1-score. We repeated all experiment 5 times and reported the means of F1-score.

Table 6 shows the results of the best performing model (i.e., RoBERTa) on our *live chat dataset*. The first row of each *live chat dataset* is the baseline, which only uses the *social media dataset* as the training set. Since we used the combinations of *social media dataset* and *live chat datasets*, training sets are represented as "Twitter + X" and "X" are the various combinations. Currently used datasets are checked in training set columns. To reduce the high similarity caused by excessive word coverage, we did not add the *social media dataset* when comparing TVcC and overlap coefficient, but only compare the similarity between each *live chat dataset*.

5 Discussion

5.1 Feasibility of Transfer Learning

Table 6 shows that in most cases, the volume of the training set is the key factor in improving performance. Moreover, Table 6 indicates that regardless of the training set used, the LC_{MC} test set achieves a relatively low F1-score. We speculate that it is because offensive data, which is only 7.2% in the entire LC_{MC} dataset, is much smaller than the *social media dataset* or even other *live chat datasets*.

Although the class distribution of the *live chat dataset* is largely different from the *social media dataset*, the *live chat dataset* achieved comparative performance for some cases. For example, on Twitter + LC_{GT} + LC_{MC} training set, we achieved 0.802 F1-score, suggesting that the transfer learning is possible.

5.2 Feasibility of TVcC

The similarity of live chat posts (TVcC) shows that high TVcC indicates high performance (F1-score) in most cases. However, with the increase in the volume of the training set, TVcC is indeed increasing, this increment seems caused by more data covering more words. If we only compare the case of adding a single training set (for instance, Twitter + LC_{GT}, Twitter + LC_{HS} or Twitter + LC_{MC} on LC_{FN} test set), we can see that F1-score has no correlation with TVcC (correlation between TVcC and F1-score is $r = 0.298$ which shows a weak correlation), suggesting high TVcC does not equal to high F1-score.

A possible reason could be that TVcC was proposed on the NER task (Dai et al. 2019). The NER task is sensitive to word representation and noisy data might not lead to highly accurate results (Mishra and Diesner 2016). However, compared to the data in the NER task, our chat data contains more user-generated contents, such as acronyms or neologisms. These contents would be considered as out of vocabulary terms by the NER systems (Mishra and Diesner 2016). Therefore, TVcC does not show a strong correlation with F1-score in our research compared to the NER task. We conclude that TVcC is not a suitable clue for our task.

5.3 Feasibility of Overlap Coefficient

Consider that only comparing a single *live chat dataset*, the similarity of target contents (overlap coefficient) shown in Table 6 indicates that F1-score shows no correlation with overlap coefficient as well as TVcC. The correlation between overlap coefficient and F1-score is $r = -0.200$ which shows a quite weak correlation. For instance, from the Table 6 we see that LC_{FN} and LC_{MC} have a high overlap coefficient. However, the F1-score is not relatively high when LC_{MC} is added to the training set.

Because game contents (keywords) may have less correlation with offensive posts, for instance, for the game GTA, as its review keywords include the words "gangster" and "shooter," it contains more violent contents than other games. However, after analyzing the content of the LC_{GT} dataset (as shown in Table 4), we found that its offensive language proportion is relatively low (12.4%). Therefore, the proposed overlap coefficient as a measure of game similarity does not indicate a strong correlation with F1-score.

5.4 Error Analysis

We manually analyzed some examples of classification errors made by our model, it still has difficulty in recognizing Internet slang and acronyms that are widely used in chat messages. A false positive example is *"i do not follow him and i am lazy so ill take your word for it."* Although this is not an offensive post, it is predicted by our model as offensive. We speculate that it is because of the word "ill" which has a negative meaning. However, "ill" in this context is an abbreviation of "I will." A false negative example is: *"if any of you are refuting that rule please stfu,"* which is an offensive post because "stfu" here is an offensive expression. However our model predicted it as not offensive.

5.5 Application

As a potential application of our research, we created an online chat room that detects offensive expressions. The RoBERTa model trained on the combination of *Twitter* + LC_{GT} + LC_{MC} sets was used for constructing the chat room. This chat room is available online at https://aoi.naist.jp/chatRoom. In order to make full use of this model, we have equipped our chat room with three modes to handle different scenarios and needs: (1) **Hide mode:** When an offensive post is detected, this mode does not send the post to the socket channel, but displays only on the sender's client as shown in Fig. 1a. (2) **Replace Mode:** When an offensive post is detected, this mode shows a modified post in which offensive expressions are masked or replaced as shown in Fig. 1b. (3) **Alert Mode:** In this mode, if offensive expressions are detected in a post, the sender is alerted before the user's post is sent into the socket channel as shown in Fig. 1c.

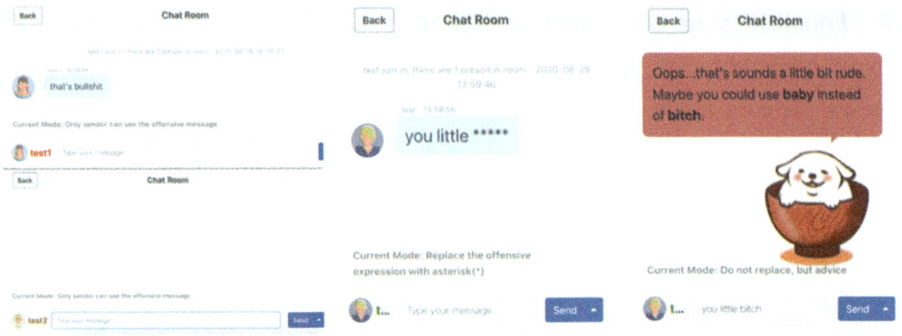

(a) Hide mode (b) Replace mode (c) Alert mode

Fig. 1. Three variations of the implementation of proposed offensive detection

6 Conclusion

This study shows that it is possible to transfer from social media to video live streaming chats on the task of offensive language detection. Moreover, we propose a solution which improves the performance of the model by combining multiple datasets when the data is limited. However, both the similarity of chat post or target content we proposed did not help to predict the transferability. One limitation of this research is that the comparison of the target (game) content similarity is quite crude.

In future work, we will collect more data on a wider variety of games. We also consider assessing the similarity of the datasets by computing the embedded vectors rather than from the surface textual information such as word occurrence that we used in this research. How to update existing training data for new data (new games) has much room for future following studies. Meanwhile, we are going to classify the targets and types of offensive language to better understand the characteristics thereof and the sender's intent.

Acknowledgements. This study was supported in part by JSPS KAKENHI Grant Number JP19K20279 and Health and Labor Sciences Research Grant Number H30-shinkougyousei-shitei-004.

References

Alakrot, A., Murray, L., Nikolov, N.S.: Towards accurate detection of offensive language in online communication in arabic. Procedia Comput. Sci. **142**, 315–320 (2018)

Baziotis, C., Pelekis, N., Doulkeridis, C.: DataStories at SemEval-2017 task 4: Deep LSTM with attention for message-level and topic-based sentiment analysis. In: Proceedings of the 11th International Workshop on Semantic Evaluation (SemEval-2017), pp. 747–754 (2017)

Bowman, S.R., Angeli, G., Potts, C., Manning, C.D.: A large annotated corpus for learning natural language inference. In: Proceedings of the 2015 Conference on Empirical Methods in Natural Language Processing, pp. 632–642 (2015)

Chapman, G.: Cranks, fetishists and monomaniacs-flamers. New Repub. **212**(15), 13–15 (1995)

Cheng, J,, Bernstein, M., Danescu-Niculescu-Mizil, C., Leskovec, J.: Anyone can become a troll: causes of trolling behavior in online discussions. In: Proceedings of the 2017 ACM Conference on Computer Supported Cooperative Work and Social Computing, CSCW 2017, pp. 1217–1230 (2017)

Cohen, J.: A coefficient of agreement for nominal scales. Educ. Psychol. Measur. **20**(1), 37–46 (1960)

Conneau, A., Schwenk, H., Barrault, L., Lecun, Y.: Very deep convolutional networks for text classification. In: Proceedings of the 15th Conference of the European Chapter of the Association for Computational Linguistics: Volume 1, Long Papers, pp. 1107–1116 (2017)

Dai, X., Karimi, S., Hachey, B., Paris, C.: Using similarity measures to select pretraining data for NER. In: Proceedings of the 2019 Conference of the North American Chapter of the Association for Computational Linguistics: Human Language Technologies, Volume 1 (Long and Short Papers), pp. 1460–1470 (2019)

Devlin, J., Chang, M.W., Lee, K., Toutanova, K.: BERT: pre-training of deep bidirectional transformers for language understanding. In: Proceedings of the 2019 Conference of the North American Chapter of the Association for Computational Linguistics: Human Language Technologies, Volume 1 (Long and Short Papers), pp. 4171–4186 (2019)

Guo, H., Pasunuru, R., Bansal, M.: Multi-source domain adaptation for text classification via distancenet-bandits. In: AAAI, pp. 7830–7838 (2020)

Joulin, A., Grave, E., Bojanowski, P., Mikolov, T.: Bag of tricks for efficient text classification. In: Proceedings of the 15th Conference of the European Chapter of the Association for Computational Linguistics: Volume 2, Short Papers, pp. 427–431 (2017)

Kingma, D.P., Ba, J.: Adam: a method for stochastic optimization. CoRR abs/1412.6980 (2014)

Liu, P., Li, W., Zou, L. NULI at SemEval-2019 task 6: transfer learning for offensive language detection using bidirectional transformers. In: Proceedings of the 13th International Workshop on Semantic Evaluation, pp. 87–91 (2019a)

Liu, Y., et al.: RoBERTa: a Robustly Optimized BERT Pretraining Approach. ArXiv abs/1907.11692 (2019b)

Matsumoto, K., Ren, F., Matsuoka, M., Yoshida, M., Kita, K.: Slang feature extraction by analysing topic change on social media. CAAI Trans. Intell. Technol. **4**(1), 64–71 (2019)

Mishra, S., Diesner, J.: Semi-supervised named entity recognition in noisy-text. In: Proceedings of the 2nd Workshop on Noisy User-generated Text (WNUT), pp. 203–212 (2016)

Nikolov, A., Radivchev, V.: Nikolov-radivchev at SemEval-2019 task 6: offensive tweet classification with BERT and ensembles. In: Proceedings of the 13th International Workshop on Semantic Evaluation, pp. 691–695 (2019)

Rezvan, M., Shekarpour, S., Alshargi, F., Thirunarayan, K., Shalin, V.L., Sheth, A.: Analyzing and learning the language for different types of harassment. PLoS ONE **15**(3), 1–22 (2020)

Waseem, Z., Davidson, T., Warmsley, D., Weber, I.: Understanding abuse: A typology of abusive language detection subtasks. In: Proceedings of the First Workshop on Abusive Language Online, pp. 78–84 (2017)

Yang, Z., Dai, Z., Yang, Y., Carbonell, J.G., Salakhutdinov, R., Le, Q.V.: Xlnet: Generalized autoregressive pretraining for language understanding. CoRR abs/1906.08237 (2019)

Zampieri, M., Malmasi, S., Nakov, P., Rosenthal, S., Farra, N., Kumar, R.: Predicting the type and target of offensive posts in social media. In: Proceedings of the 2019 Conference of the North American Chapter of the Association for Computational Linguistics: Human Language Technologies, Volume 1 (Long and Short Papers), Association for Computational Linguistics, pp. 1415–1420 (2019a)

Zampieri, M., Malmasi, S., Nakov, P., Rosenthal, S., Farra, N., Kumar, R.: SemEval-2019 task 6: identifying and categorizing offensive language in social media (OffensEval). In: Proceedings of the 13th International Workshop on Semantic Evaluation, pp. 75–86 (2019b)

Zhang, X., Zhao, J., LeCun, Y.: Character-level convolutional networks for text classification. In: Proceedings of the 28th International Conference on Neural Information Processing Systems - Volume 1, NIPS 2015, pp. 649–657 (2015)

BERT-Based Dialogue Evaluation Methods with RUBER Framework

Khin Thet Htar[1(✉)], Yanan Wang[2], Jianming Wu[2], Gen Hattori[2], and Aye Thida[1]

[1] University of Computer Studies, Mandalay, Myanmar
{khinthethtar,ayethida}@ucsm.edu.mm
[2] KDDI Research, Inc., Fujimino, Japan
{wa-yanan,ji-wu,ge-hattori}@kddi-research.jp

Abstract. Dialogue systems are embedded in smartphones and Artificial Intelligence (AI) speakers and are widely used through text and speech. To achieve a human-like dialogue system, one of the challenges is to have a standard automatic evaluation metric. Existing metrics like BLEU, METEOR, and ROUGE have been proposed to evaluate dialogue system. However, those methods are biased and correlate very poorly with human judgements of response quality. On the other hand, RUBER is applied to not only train the relatedness between the dialogue system generated reply and given query, but also measure the similarity between the ground truth and generated reply. It showed higher correlation with human judgements than BLEU and ROUGE. Based on RUBER, instead of static embedding, we explore using BERT contextualised word embedding to get a better evaluation metrics. The experiment shows that our evaluation metrics using BERT are more correlated to human judgement than RUBER. Experimental results show that BERT feature based evaluation metric had 0.31 and 0.26 points higher scores and BERT fine tune evaluation metric got higher 0.39 and 0.36 points in Pearson and Spearman correlation with human judgement score than RUBER, respectively.

Keywords: Evaluation metric · Dialogue system · Contextualised embedding · BERT

1 Introduction

A dialogue system, also known as a conversational agent, is a program to communicate with humans automatically in natural language (text, speech, or even both). The human-like conversational agents, chatbots, are getting more prominent and are applied in a wide range of areas. They are designed to mimic unstructured human-like conversations rather than focusing on a particular task like booking hotel or giving travel direction. In developing dialogue systems to achieve more human likeness, many researchers resort to manual annotation to evaluate their models. It is crucial to have an accurate automatic evaluation metric in dialogue system research in order to allow rapid testing of new models with less expense and higher scalability than human evaluation.

K. Yada et al. (Eds.): JSAI 2020, AISC 1357, pp. 133–144, 2021.
https://doi.org/10.1007/978-3-030-73113-7_12

Existing evaluation metrics, PARADISE [1] and MeMo [2] used for task based dialogue system, are not appropriate for open domain dialogue system that handled diversity and uncertainty of utterances. Although BLEU [3] and METEOR [4] that have been used for machine translation and ROUGE [5] that has been used for automatic summarisation are adopted for evaluation of dialogue system, they correlate poorly with human judgement [6] because they can capture only lexically word overlapping. In recent works [7,8], embedding based evaluation metrics were introduced considering meaning of each word in query and reply. But they are based only on static embedding. Although static embedding has a fixed context independent representation for each word, it cannot represent the rich semantics of words in contexts.

This paper explores the efficiency of contextualised Bidirectional Encoder Representations from Transformers(BERT) [9] word embedding by conducting under RUBER [8] referenced and unreferenced scorer framework. BERT pretrained model is used not only in measuring the semantic textual similarity between groundtruth and generated reply but also in predicting the relatedness between given query and generated reply. The relatedness is measured by two approaches, BERT feature based and fine tune. The feature based approach uses BERT pretrained vector representations for training network and the fine tune approach is trained by simply fine tuning all BERT pretrained parameters on query-reply pairs of DailyDialog [10] dataset.

We evaluated proposed method on attention based dialogue system. According to Pearson and Spearman correlation, the experiment shows that proposed BERT feature based and BERT fine tune evaluation metrics are more correlated to human judgement than RUBER.

2 Related Works

In contrast to the traditional dialogue evaluation methods, such as BLEU and ROUGE, which are computed using generated and ground-truth replies, some works have been attempted to train models on query-reply pairs as dialogue evaluation methods.

The ADEM [7], neural network based evaluation metric, is introduced to predict appropriateness of the system replies for the given queries or utterances. ADEM predicts highly correlated scores with human judgment scores than BLEU and ROUGE. But in order to train the model, it requires human annotated score for appropriateness of each query-reply pair.

Unsupervised evaluation metric named Referenced metric and Unreferenced metric Blended Evaluation Routine (RUBER) [8] was proposed for open-domain dialogue systems. RUBER evaluates reply generated from dialogue system by blending the score of referenced and unreferenced metrics. In referenced metric, the closeness of ground truth and generated reply is measured by cosine similarity between their pretrained word embeddings. In unreferenced metric, the relatedness that represents correlation between the generated reply and its query is measured. Both referenced and unreferenced metrics depend on static word embedding that does not capture contextualized information.

Recently, BERT contextualized word embedding was proposed that learns words' representations from both left and right context by using a "masked language model" (MLM) pretraining objective and jointly pretrains text-pair representations by using a "next sentence prediction" task. Powerful BERT outperforms many task-specific architectures, achieving state-of-the-art performance on sentence level and token level tasks such as Semantic Textual Similarity Benchmark(STS-B) [9], Multi-Genre Natural Language Inference(MNLI) [11], translation system [12] etc. BERT is a method that can be used to extract word embedding features that can properly represent the rich context semantics of words. Therefore, aiming to achieve a dialogue evaluation method that is much more correlated with human judgment than RUBER, we build referenced metric using BERT-based embedding instead of static word embedding, and model BERT feature based and fine tune unreferenced metrics instead of RUBER' counterpart.

3 Method

The overall design of proposed method is conducted under RUBER framework by blending sentence-BERT referenced metric with two different unreferenced metrics, BERT feature based unreferenced metric and BERT fine tune unreferenced metric, separately. Blending of referenced and unreferenced metrics is performed by different heuristics. The final dialogue evaluation score is computed based on the blending approach.

3.1 Sentence-BERT Referenced Metric

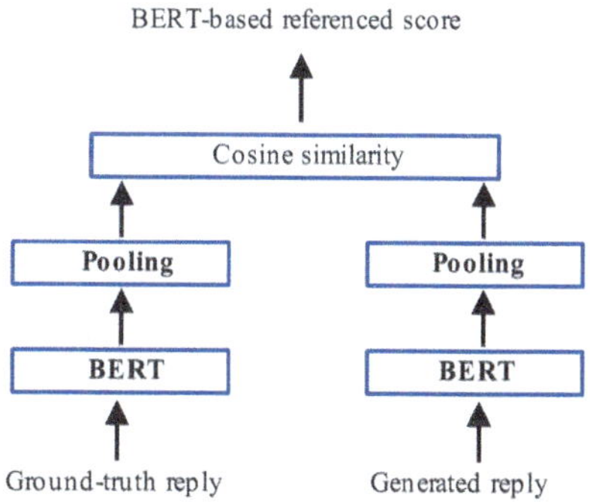

Fig. 1. Sentence-BERT referenced metric

Referenced metric is constructed to measure the similarity between the referenced groundtruth reply and generated reply from attention based dialogue system which is introduced in 4.1. Sentence BERT model is utilized that fine tunes BERT model with a siamese or triplet network structure [13] to produce semantically meaningful sentence embeddings for semantic textual similarity by adopting simple pooling strategy. These sentence embeddings generated from sentence BERT are used for similarity measure between groundtruth and generated replies via cosine similarity as shown in Fig. 1.

3.2 Unreferenced Metrics

In evaluation of dialogue system, it is also important to think about how related the reply is to the given query. Like RUBER, unreferenced metric is designed to measure the relatedness between query and dialogue model generated reply. Two different unreferenced methods, BERT feature based unreferenced metric and BERT fine tune unreferenced metric, were proposed. These two metrics are based on BERT pretrained model.

BERT Feature Based Unreferenced Metric. In the proposed BERT feature based unreferenced metric, each word in query and reply is encoded into vector using pretrained BERT word embedding as shown in Fig. 2. Since BERT embeddings include complete information about word's context by training on bidirectional transformer, simple vector pooling approach is adopted to summarise sentence information with fixed-size sentence embedding instead of Bi-GRU-RNN used in RUBER. The vector representations of query and reply is concatenated with learning parameter which is calculated by multiplying transpose of query (q^T), parameter matrix (M) and reply (r). The concatenation is fed into multilayer perceptron (MLP) to predict relatedness or unreferenced score.

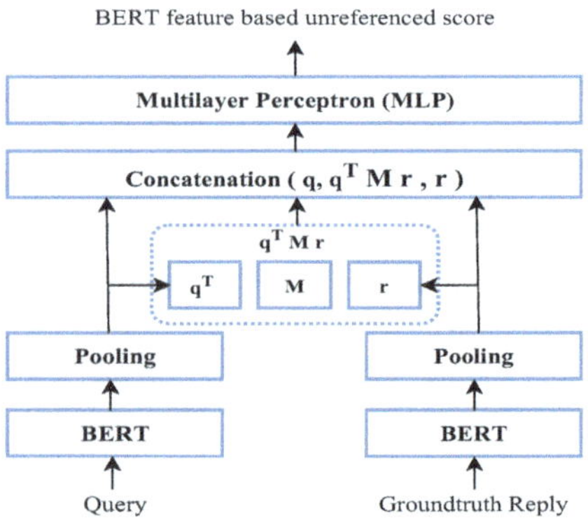

Fig. 2. BERT feature based unreferenced metric

Learning process is performed by adopting negative sampling approach which does not require human label in loss calculation. For each groundtruth query-reply sample (q,r), another reply in the training data set is randomly paired with query (q) as negative sample (q,r^-).

$$Loss = max\{0, \Delta + (-1 * S_U(q,r)) + (1 * S_U(q,r^-))\} \tag{1}$$

As RUBER, loss function is calculated by predicted feature based unreferenced score(S_U) of positive sample (q,r) and predicted feature based unreferenced score(S_U) of randomly selected negative sample (q,r^-) as described in Eq. 1. Δ is set a constant value(0.5) .

BERT Fine Tune Unreferenced Metric. BERT paper shows a number of ways for BERT fine tune on different tasks including sentence pair classification tasks which are similar to relatedness measure of query-reply pairs. We fine tune BERT pretrained model to achieve unreferenced score. By incorporating with BERT pretrained model, fine tuning is started by adding one extra output layer and the network is trained on the query-reply pairs for relatedness measure.

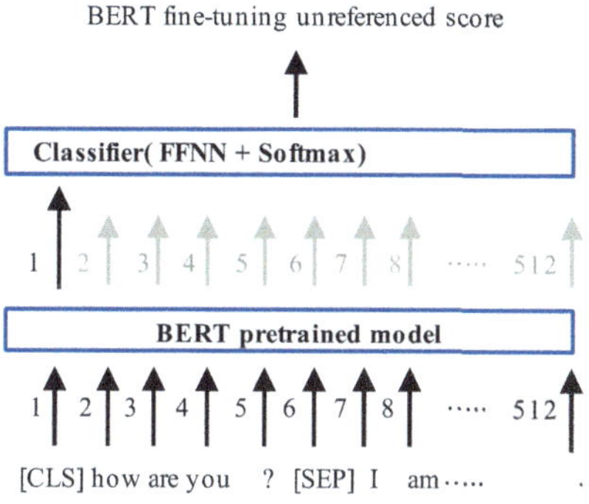

Fig. 3. BERT fine tune unreferenced metric

Due to the data are needed to be supervised in fine tuning process, negative sampling approach, which is adopted by some researchers for utterance matching [14], is used to add label for each query-reply pair in the training dataset. In negative sampling, all queries in training dataset are randomly paired with replies and labeled 0 indicating negative relatedness. And all original query-reply pairs in training dataset are labeled as 1 indicating positive relatedness.

Before fine tuning, [CLS] token is inserted at the beginning of query and a [SEP] token is inserted at the end of query and reply for each training query-reply pair. These preprocessed query-reply pairs with relatedness label were plugged into BERT pretrained model as shown in Fig. 3 and fine tune all the parameters end-to-end. BERT vector representation of [CLS] token, which summarises sentence information for classification tasks, is fed into an output feed forward neural network layer to learn parameters for relatedness measure. And loss function is calculated using cross entropy loss which takes groundtruth relatedness score, y and the estimated value of each query-reply pair, $S_U(q,r)$, generated from the last classification layer as described in Eq. 2.

$$Loss = -y \ log(\ S_U(q,r) \ ) \tag{2}$$

3.3 Blending Approach

Referenced and unreferenced metrics are combined by using simple heuristics such as min, max, geometric mean and arithmetic mean in order to calculate evaluation score. Before combination of metrics, score predicted by each metric is normalized to same scale range between 0 and 1.

4 Experiment

We compare proposed dialogue evaluation method with baseline RUBER to prove that proposed methods have better correlation to human judgement. In this section, considering that the dialogue evaluation method is built and evaluated based on the dialogue system, we firstly describe the attention-based dialogue system that is constructed to acquire generated reply for evaluation, and then introduce the DailyDialog corpus that is used for training unreferenced metrics. Finally, preparation for human judgement is explained and the human correlation with evaluation metrics is analyzed and discussed.

4.1 Attention Based Dialogue System

The attention-based dialogue system is applied to generate replies required for building proposed dialogue evaluation method. This dialogue system is constructed by using two separate recurrent neural networks together with an attention-based encoder-decoder architecture [15]. One RNN acts an encoder, which captures semantic information of given query by encoding a variable length input query sequence to a fixed-length context vector (final hidden layer of encoder). The second RNN is a decoder, which guess what the next word will be in the sequence and a hidden state by taking the input word and context vector. Next iterations are performed based on previous predicted word and hidden state. The decoder uses an "attention mechanism" that pay attention to certain parts of the input sequence, rather than using the entire fixed context at every step. This dialogue system is trained on movie scripts from Cornell-Movie Dialog Corpus [16]. The Cornell Movie-Dialogs Corpus is a large and diverse dataset which includes 304,713 total utterances from conversational exchanges between 10,292 pairs of movie characters.

Table 1. Data statistics for unreferenced metrics

Dataset	Dialogues (number)	Query-reply pairs (number)
Train	11,118	76,052
Develop	1,000	7,069
Test	1,000	3,370

4.2 Data Description

Different copora are used for constructing dialogue system and evaluation metrics to be able to handle diversity. We train evaluation metrics utilizing DailyDialog corpus [10] instead of the Cornell Movie Dialogs corpus used for dialogue system training. The DailyDialog corpus covers various topics about our daily life, and contains less noise such as typos and slangs than Ubuntu Corpus [17,18] and Twitter Dialog Corpus [19]. As shown in Table 1, this corpus is comprised of 11118, 1000 and 1000 multi-turn dialogues for train, develop and test sets, respectively and each dialogue includes 7.9 turns in average.

In order to train unreferenced metrics, query-reply pairs are extracted from dialogues included in training, developing and testing sets of DailyDialog corpus, respectively as shown in Table 1.

For correlation analysis between human judgement score and predicted score of evaluation metrics, 300 query-reply pairs are randomly selected from testing dataset of DailyDialog.

Table 2. Example of query-reply pairs that are rated by three volunteers with score from 1–5 (not appropriate to completely appropriate)

Query	Reply	Human judgement
May I know where you are going?	I am going to california	5 , 5 , 5
Should we tip? This service is pretty good	It's just a good idea	2 , 4 , 5
Sure you can; when would you like to come in?	I ll be in a minute	5 , 4 , 2
What's she doing?	She is praying	5 , 4 , 5

4.3 Human Judgement

In term of human judgement process, volunteers were invited to make judgement for replies generated by attention based dialogue system. Each volunteer was provided 100 query-reply pairs with two extra pairs for attention checking and asked to rate each pair based on the appropriateness of the reply for the given query on a scale of 1–5 (not appropriate to very appropriate). We removed all pairs that were rated by volunteers who failed to correctly answer attention check

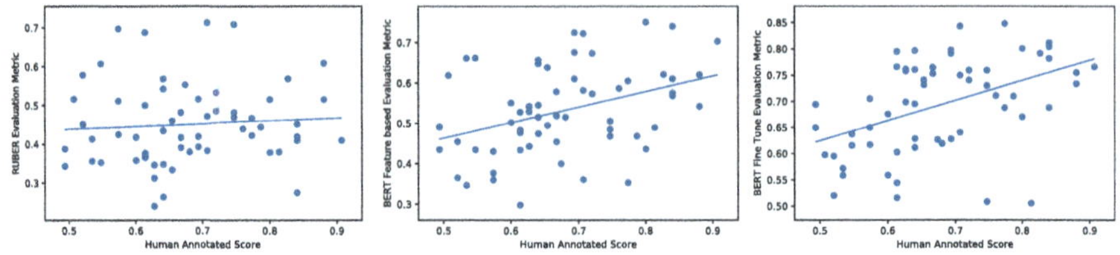

Fig. 4. Scatter plots showing the correlation between the evaluation metrics and human annotated scores

tests. For 300 pairs of testing, each pair was annotated by 3 volunteers and taken average annotated score. And we used Fleiss Kappa [20] for the measurement of agreement among the raters in judging relatedness of query-reply pairs. The overall inter-annotator agreement scores for questions asked in the survey got 0.33 which indicates fair agreement [21]. Table 2 demonstrates four query-reply pairs with their human annotated scores rated by three volunteers.

4.4 Result

The proposed method is compared with baseline RUBER. BERT feature based evaluation metric blended Sentence-BERT referenced metric and BERT feature based unreferenced metric. BERT fine tune evaluation metric is the combination of Sentence-BERT referenced metric and BERT fine tune unreferenced metric.

Table 3. Comparison results of evaluation metrics and human score

Query	Ground-truth reply	Generated reply	Human score	RUBER	BERT feature based	BERT fine tune
Sure you can; when would you like to come in?	I'd like to come in tomorrow morning at 9	I ll be in a minute .	0.7333	0.4658	0.8136	0.8141
Who gave my dog this cooked chicken bone ?	I did. Why?	My husband . . .	1.0000	0.7247	0.5873	0.8256
What's the problem?	I wrote the check out for way too much.	I can t be here . .	0.6667	0.3921	0.6161	0.6792
Average of 300 samples	–	–	0.6777	0.4533	0.5330	0.6766

We randomly select three samples used in human judgement. Table 3 shows human score, the evaluation results of RUBER, BERT feature based and BERT fine tune evaluation metrics. Human annotated scores are normalized to the range of 0 and 1 to compare with score of different evaluation scores. But it does not affect to the degree of correlation. In blending process, arithmetic mean is used here. Compared to RUBER, BERT feature based and BERT fine tune evaluation metrics are more similar to human score by taking average of the 300 samples.

Scatter plots depict linear relationship between system evaluation score and human annotated score. As shown in Fig. 4, the second and third scatter plots, BERT feature based and BERT fine tune evaluation metrics have stronger linear relationship than RUBER. This means score predicted by proposed metrics are more correlated to human annotated score for attention based dialogue system.

Furthermore, in order to analyze the correlation between human score and evaluation metrics from linear and monotonic relationships, we calculated Pearson and Spearman correlations [8]. Pearson correlation can evaluate linear relationships and Spearman correlation can evaluate monotonic relationships between human score and predicted score of evaluation metrics. A relationship is linear when a change in human annotated score is associated with a proportional change in evaluation metric's score. A relationship is monotonic when scores tend to change together but it is not in constant rate. Table 4 shows Pearson and Spearman correlation scores between human judgement and three different evaluation metrics, RUBER, BERT feature based and BERT fine tune evaluation metrics with different blending heuristics strategies such as min, max, geometric mean and arithmetic mean and also describes correlation scores of their respective referenced and unreferenced metrics without blending.

Table 4. Correlation between evaluation metrics and human annotation. *Min, Max, Geometric mean and Arithmetic mean* indicate blending approaches.

Evaluation methods		Pearson	Spearman
RUBER	Referenced	−0.1403	−0.1462
	Unreferenced	**0.1663**	0.1475
	Min	0.1325	0.1421
	Max	−0.0463	−0.0342
	Geometric mean	0.1606	**0.1919**
	Arithmetic mean	0.0667	0.1321
BERT feature based	Referenced	0.2488	0.2692
	Unreferenced	0.3161	0.3727
	Min	0.3532	0.3815
	Max	0.3434	0.354
	Geometric mean	**0.3938**	**0.4161**
	Arithmetic mean	0.3759	0.3901
BERT fine tune	Referenced	0.2488	0.2692
	Unreferenced	0.4352	**0.5064**
	Min	0.4015	0.4055
	Max	0.4506	0.4916
	Geometric mean	0.4279	0.4477
	Arithmetic mean	**0.4535**	0.4877

It is obvious that BERT fine tune evaluation metric has the highest Pearson score by blending referenced and unreferenced metrics with arithmetic mean and highest Spearman score by BERT fine tune unreferenced metric without blending with referenced metric. We can also notice that the unreferenced metric has much higher Pearson and Spearman correlation scores than referenced metric. In comparing just referenced metrics before blending, BERT based referenced metric had better Pearson and Spearman correlation scores than referenced metric of baseline RUBER. Although BERT feature based is not as good as BERT fine tune correlation score, it got much higher Pearson and Spearman correlation score than baseline RUBER and geometric mean blending strategy strikes best correlated score in BERT feature based evaluation metric.

4.5 Discussion

According to the correlation analysis results in 4.4, BERT-based methods (BERT feature based, BERT fine tune) correlate with human score much more than RUBER, since BERT-based embeddings can properly represent the rich context semantics of words compared to static word embedding. On the other hand, unreferenced metric has higher correlation scores than referenced metric, we believe that relationships between dialogue contexts such as query-reply pairs are important to consider when evaluating dialogue system, and thus modeling these relationships can contribute to the achievement of human-like dialogue evaluation methods. If referenced metric is also taken into account, it is needed to choose good blending strategy which can give better human correlation score. And negative sampling approach can also be applied effectively in BERT fine tuning which requires relatedness label for each query-reply pair in training metrics because it needs supervised data. In particular, our proposed BERT fine tune evaluation metric shows the highest correlation scores with human judgement than other evaluation metrics, BERT feature based and baseline RUBER. We can conclude that fine tuning on BERT pretrained parameters is effective way to achieve better unreferenced metric that is used to predict relatedness of a reply generated from the dialogue system for a given query.

5 Conclusion

In this paper, we proposed BERT-based dialogue evaluation methods with RUBER framework. Instead of static word embedding, we used BERT contextualized word embedding. We built Sentence-BERT referenced metric based on SBERT model, and used BERT pretrained model to construct BERT feature based unreferenced metric and fine tuned to achieve BERT fine tune unreferenced metric. By blending referenced and unreferenced metrics, BERT fine tune evaluation metric, combination of Sentence-BERT referenced metric and BERT fine tune unreferenced metric, is most correlated to human score than others. In the future, we would like to construct more human correlated evaluation metric by considering topicality among multi-turn conversations to be effectively applied in dialogue system research.

References

1. Walker, Marilyn A., et al.: PARADISE: a framework for evaluating spoken dialogue agents. In: 35th Annual Meeting of the Association for Computational Linguistics and 8th Conference of the European Chapter of the Association for Computational Linguistics, Association for Computational Linguistics (1997)
2. Möller, S., et al.: MeMo: towards automatic usability evaluation of spoken dialogue services by user error simulations. In: Ninth International Conference on Spoken Language Processing (2006)
3. Papineni, K., Roukos, S., Ward, T., Zhu, W.-J.: BLEU: a method for automatic evaluation of machine translation. In: Proceedings of the 40th annual meeting on association for computational linguistics. Association for Computational Linguistics, pp. 311–318 (2002)
4. Banerjee, S., Lavie, A.: METEOR : an automatic metric for MT evaluation with improved correlation with human judgements. In: Proceedings of the Workshop on Intrinsic and Extrinsic Evaluation Measures for Machine Translation and/or Summarisation (2005)
5. Lin, C.-Y.: ROUGE : a package for automatic evaluation of summaries. In: Text Summarisation Branches out : Proceedings of the ACL-04 Workshop (2004)
6. Liu, C.-W., Lowe, R., Serban, I., Noseworthy, M., Charlin, L., Pineau, J.: How NOT to evaluate your dialogue system : an empirical study of unsupervised evaluation metrics for dialogue response generation. In: Proceedings of the 2016 Conference on Empirical Methods in Natural Language Processing (2016)
7. Lowe, R., Noseworthy, M., Serban, I.V., Angerlard-Gontier, N., Bengio, Y., Pineau, J.: Towards an automatic turing test : learning to evaluate dialogue responses. In: Proceedings of the 55th Annual Meeting of the Association for Computational Linguistics, Association for Computational Linguistics, pp. 1116–1126 (2017)
8. Tao, C., Mou, L., Zhao, D., Yan, R.: RUBER : an unsupervised method for automatic evaluation of open-domain dialog systems. In: Mcllraith and Weinberger (2018)
9. Devlin, J., Chang, M.-W., Lee, K., Toutanova, K.: BERT : pre-training of deep bidirectional transformers for language understanding. In: Proceedings of NAACL-HLT, pp. 4171–4186 (2018)
10. Li, Y., et al.: Dailydialog: A manually labelled multi-turn dialogue dataset. In: Proceedings of the Eighth International Joint Conference on Natural Language Processing (Volume 1: Long Papers) (2017)
11. Williams, A., Nangia, N., Bowman, S.: A broad-coverage challenge corpus for sentence understanding through inference. In: Proceedings of the 2018 Conference of the North American Chapter of the Association for Computational Linguistics: Human Language Technologies, Volume 1 (Long Papers), pp. 1112–1122 (2018)
12. Imamura, K., Sumita, E.: Recycling a pre-trained bert encoder for neural machine translation. In: Proceedings of the 3rd Workshop on Neural Generation and Translation (WNGT 2019), pp. 23–31 (2019)
13. Reimers, N., Gurevich, I.: Sentence-bert : Sentence embeddings using siamese bert-networks. In: Proceedings of the 2019 Conference on Empirical Methods in Natural Language Processing and the 9th International Joint Conference on Natural Language Processing, Association for Computational Linguistics (2019)
14. Yan, R., Song, Y., Wu, H.: Learning to respond with deep neural networks for retrieval-based human-computer conversation system. In: Proceedings of the 39th International CM SIGIR conference on Research and Development in Information Retrieval. pp. 55–64 (2016)

15. Luong, T., Pham, H., Manning, C.D.: Effective approaches to attention-based neural machine translation. In: Proceedings of the 2015 Conference on Empirical Methods in Natural Language Processing, Association for Computational Linguistics (2015)
16. Danescu-Niculescu-Mizil, C., Lee, L.: Chameleons in imagined conversations : a new approach to understanding coordination of linguistic style in dialogs. In: Proceedings of the 2nd workshop on cognitive modelling and computational linguistics, Association for Computational Linguistics (2011)
17. Lowe, R., Pow, N., Serban, I., Pineau, J.: The Ubuntu dialogue corpus: a large dataset for research in unstructured multi-turn dialogue systems. In: Proceedings of the 16th Annual Meeting of the Special Interest Group on Discourse and Dialogue, pp. 285–294, Prague, Czech Republic. Association for Computational Linguistics (2015)
18. Zhao, T., Pavel, A., Eskenazi, M., Gupta, P., Mehri, S., Bigham, J.P.: Investigating evaluation of open-domain dialogue systems with human generated multiple references. In: Proceedings of the SIGDial 2019 Conference, pp. 379–391 (2019)
19. Ritter, A., Cherry, C., Dolan, W.B.: Data-driven response generation in social media. In: Proceedings of the 2011 Conference on Empirical Methods in Natural Language Processing, pp. 583–593, Edinburgh, Scotland, UK. Association for Computational Linguistics (2011)
20. Fleiss, J.L.: Measuring nominal scale agreement among many raters. Psychol. Bull. **76**, 378–382 (1971)
21. Landis, J., Koch, G.: The measurement of observer agreement for categorical data. Biometrics **33**(1), 159–74 (1977)

The Morandi Room: Entering the World of Morandi's Paintings Through Machine Learning

Shigeru Kobayashi[1](✉), Ryota Kuwakubo[1], Shigeru Matsui[1], Yoshiyuki Otani[1], Xinqi Zhang[1], and Daisuke Niizumi[2]

[1] Institute of Advanced Media Arts and Sciences, Ogaki, Gifu 503-0006, Japan
mayfair@iamas.ac.jp
[2] daisukelab, Tokyo, Japan
https://www.iamas.ac.jp/en/

Abstract. In this study, we propose a new way to appreciate artwork based on the growing interest in the active appreciation of artwork and the development of machine learning technology. We chose Italian painter Giorgio Morandi, who was active in the first half of the 20 century and known for his unique composition and coloring, as the theme, and developed a hands-on exhibit in which spectators freely arrange and compose objects that reproduce the painter's motifs, and generates images that reproduce the painter's coloring by machine learning. From the results of a questionnaire survey of 29 people, we confirmed that experiencing this exhibit deepened their interest in the painter.

Keywords: Machine learning · Art · Education · Art appreciation · Creativity

1 Introduction and Related Work

In 1950, Alan M. Turing proposed the question, "Can machines think?" in his paper "Computing Machinery and Intelligence" [15]. Art has long been a hot topic when thinking about machine intelligence, ever since the later-known "Turing Test" first asked "Please write me a sonnet on the subject of the Forth Bridge" in the last century.

Many projects have concerns to apply machine learning to the visual arts; for example, the Next Rembrandt project[1]—conducted in 2016 by ING, Microsoft, TU Delft, and the Mauritshuis museum—analyzed the surviving works of the 17th-century painter Rembrandt in details by experts and machine learning, and created a "new work" with motifs, compositions, and touches which were presumed as typical of Rembrandt. Another example is DeepArt[2]—run by

[1] https://www.nextrembrandt.com/.
[2] https://deepart.io/.

K. Yada et al. (Eds.): JSAI 2020, AISC 1357, pp. 145–156, 2021.
https://doi.org/10.1007/978-3-030-73113-7_13

researchers in Germany, Switzerland, and Belgium—which has offered a web service since 2016 that leverages the algorithm created by Leon Gatys [4] enabling users to easily create a new "artwork" just by simply uploading a photo and choosing a style. The recent example by the artist and programmer Gene Kogan is "Abraham", which aims to build an agent that autonomously generates unique and original art since 2019 [7]. Furthermore, Ahmed M. Elgammal et al. proposed the concept of "Creative Adversarial Networks," which learns established styles and produces works which are considered creative by deviating moderately [3].

Other fields besides visual arts include music, such as the generation of new music by machine learning [1]. In response to this situation, Glenn W. Smith and Frederic F. Leymarie edited a special issue of *Arts* entitled "The Machine as Artist," pointing out that we can begin to think the machine as a creator or collaborator, instead of an artist's subject or medium [14]. Thus, art has been an essential theme in the history of machine intelligence research and it continues to be explored by many researchers and artists from divergent perspectives.

The essential activity in art is not only in its production, but also in its appreciation. In school education, the appreciation of artwork is also considered to help improve the student's thinking and communicating abilities among a wide range of age groups. Visual Thinking Strategies, proposed in the 1980s by cognitive psychologists Abigail Housen and Philip Yenawine—the former Director of Education at the Museum of Modern Art in New York—emphasized the ability of spectators sharing their impressions of a work of art, and believed to foster the observing, critical thinking, and communicating abilities [16]. Therefore, the method is currently adopted in many museums. Just as the purpose of the experience of appreciation in art education has expanded from unilaterally conveying the author's background and historical considerations of art, so has its methods.

For example, in addition to conventional methods such as displaying an original work at a museum or projecting an image of the work with a projector, new methods using various technologies have been proposed. Some implementations include projecting images onto a large wall[3], displaying in a VR space [8], and displaying with information using AR technology [13]. As we have seen, research utilizing machine learning has focused mainly on the production of artworks. On the other hand, the importance of experiencing artworks has been noticed, and various methods have been proposed. This study builds upon this fundamental and attempts to create a new way of experiencing the appreciation of artwork utilizing machine learning.

We chose Giorgio Morandi—an Italian painter known as one of the foremost painters of the 20th century—as the subject of this study. Morandi produced more than 1,300 oil paintings over about 50 years, which most of these are still

[3] For example, the event "Immersive Museum" which was scheduled to be held in Tokyo from April 2020, projects images of works by impressionist artists such as Degas and Renoir over the entire field of vision so people can enter the world of painting. https://immersive-museum.jp/.

lifes [11]. His works are characterized by its unique composition; its coloring, its distortion of contours, and its positioning of objects which are gathered together rather than isolated (see Fig. 1[4]). Not only are his works appreciated by many artists but often becomes the subject of advanced research efforts.

Fig. 1. Giorgio Morandi, Still Life (1946)

House of ZKA—a collection of architectural investigations—held a workshop "Morandi-esque" in 2018 to create 3D models by tracing distortions in the contours of objects by Morandi as a collection of arcs and printed with a 3D printer to exhibit[5]. Guido Salimbeni et al. presented a system that supports the placement of objects in 3D space by suggesting arrangements modeled on Morandi's work [12]. Many workshops have been held in conjunction with exhibitions in order to gain a deeper understanding of Morandi, by simply reproducing his works. For example, a workshop "dissect Morandi's technique"—held in April 2016 by painter Yoichi Miyajima and others as an event related to an exhibition held in Tokyo in 2016—reproduced objects and tables appearing in Morandi's works in full size, and verified viewpoints and lighting using a digital camera and lighting equipment[6]. Inspired by these workshops, our study presents a hands-on exhibit utilizing machine learning to motivate spectators to deeply understand the artist and the world of his work.

2 Implementation

2.1 Concept and Overview of the Exhibit

"The Morandi Room" is a hands-on exhibit that proposes a new way of appreciating artwork in which spectators try to reconstruct a part of the world of the

[4] ©DACS, 2020, Photo ©Tate, CC-BY-NC-ND 3.0 (Unported) https://www.tate.org.uk/art/artworks/morandi-still-life-n05782.
[5] http://www.houseofzka.com/morandi-esque.
[6] http://fuminaosuenaga.com/workshop/morandi/.

artist's work themselves by arranging objects imitating the main motifs utilized in Morandi's work, and appreciating the world through "eyes" using machine learning models derived from Morandi's surviving paintings. Figure 2 shows the overall system, and Fig. 3 shows a visitor experiencing the exhibit. The entire exhibition space is W 3,960 mm × D 2,655 mm × H 2,250 mm, and a table W 910 mm × D 845 mm × H 810 mm is placed in it. The image of objects (up to nine objects of four types) on the table is captured by a digital camera at a distance of about 2,000 mm and a height of about 1,700 mm and is output by HDMI with 1,280 × 720 pixels at 60 fps. The output is captured in an HMDI capture box connected to a PC with an NVIDIA GeForce GTX 1080, the center 512 pixel square of the frame is cropped, and given to a machine learning model for inference to generate an image. The resulting image is then upscaled using bilinear interpolation to 1,024 pixel squares and output to a 300 mm square monitor embedded in the frame. In front of the exhibition space, we placed books related to Morandi that visitors could freely view.

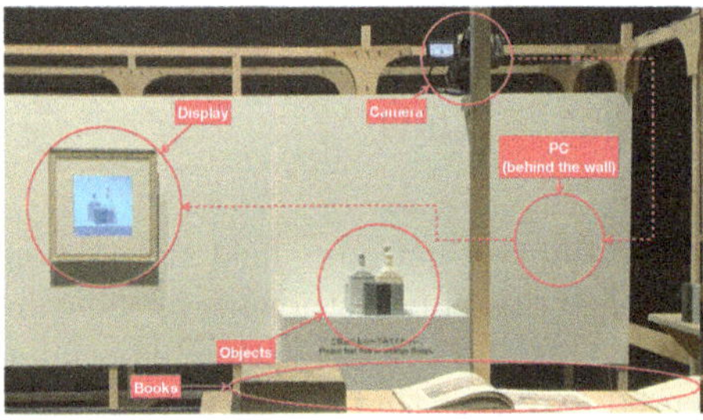

Fig. 2. System

Fig. 3. A visitor experiencing the exhibit

2.2 Dataset and Algorithm

Among the characteristics of Morandi's works, we left the composition to the spectator and handled the reproduction of coloring by machine learning models learned from Morandi's works. For the images of Morandi's paintings, we collected what appeared to be still-life paintings by Morandi on the Internet, and prepared a data set consisting of 179 images that were confirmed using Morandi's catalog. When it was not very easy to determine the position of objects because they were similar, we utilized Morandi's signature and position to identify them.

We mainly evaluated two algorithms, pix2pix by Phillip Isola et al. [6], and CycleGAN by Jun-Yan Zhu and Taesung Park et al. [17]. We also examined the colorization proposed by Aditya Deshpande et al. [2] and the style conversion proposed by Leon A. Gatys et al. [5] However, in the latter two cases, the context of the captured images remained strong, and the parts learned from Morandi's images became weak, so we focused on the first two for building our exhibit.

pix2pix is an algorithm that allows conversion in both directions from A to B or B to A during inference by providing pairs of corresponding images from domain A and B during learning. In this exhibit, we searched for intermediate representation that can handle different inputs in common, such as images of artworks actually drawn by Morandi during learning, and images of objects that reproduces the motif used by Morandi during inference. We compared the Canny edge detection only, a color-based segmentation using k-means clustering only, and edge detection and segmentation, and chose the combination of both because obtained the best results (see Fig. 4).

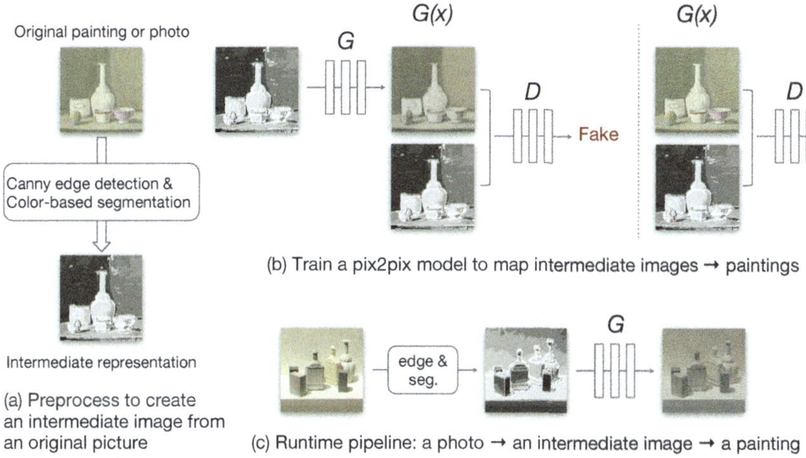

Fig. 4. pix2pix

CycleGAN is an algorithm that provides domain A and domain B images without pairing during learning and allows conversion in both directions from A to B or B to A during inference. During learning, the images taken in the exhibition environment are given as domain A, and the images of the paintings drawn by Morandi are given as domain B. During inference, the camera images are given as input, and the images are generated end-to-end (see Fig. 5). As for the images to be used as training data for CycleGAN, we placed objects on the table where environmental conditions such as lighting were determined at the exhibition site and took 51 variations. In the training of CycleGAN, the trend of the output images differed significantly from epoch to epoch during training. While some were quite close to Morandi's style of painting, some were completely different, and there was no tendency for them to become more stable with each epoch. For this reason, we adopted the model with the best epoch based on subjective judgment among the 300 epochs tested.

Subjective evaluation of the outputs of the two algorithms for the test data showed that pix2pix had a less overall failure rate and provided more stable results, whereas CycleGAN had a more significant difference in quality depending

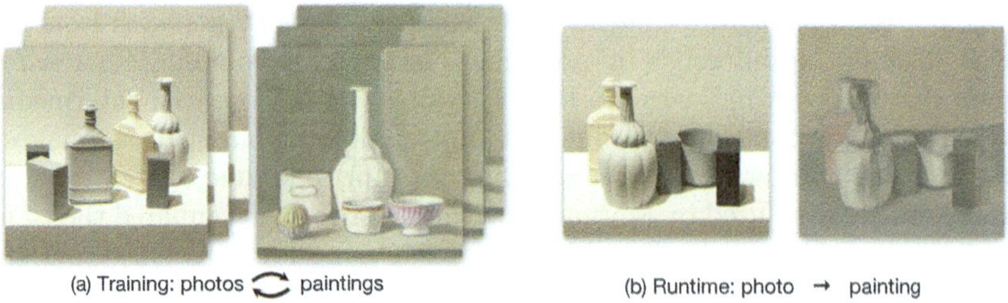

Fig. 5. CycleGAN

Fig. 6. The progression of learning and changes in the results generated by CycleGAN (epoch = 100, 150, 200, 250, 300)

on placement. Therefore, we chose pix2pix as the default algorithm and switched to CycleGAN in response to questions from spectators. On a PC with an NVIDIA GeForce GTX 1080, the frame rate was about 0.9 fps on pix2pix and about ten fps on CycleGAN.

2.3 3D Models

We modeled the objects to be placed on the table in full scale based on the photo album "Morandi's Objects" by photographer Joel Meyerowitz, featuring the object used as a motif by Morandi [9], and dimensional data obtained from the Morandi Museum. We created 3D models of objects with complex shapes like pots or bottles, printed them out using a 3D printer, and painted them with acrylic paint. For objects with simple shapes, such as cubes, we shaped them out of cardboard and colored them with acrylic paint.

3 Results

3.1 Example of Arrangements by Visitors

We presented this exhibit at a four-day exhibition. The visitors during the four days generated 789 unique patterns as shown in Fig. 7. Some seemed to be conscious of Morandi's original work. Others tried unusual arrangements, such as by turning objects upside down. Additionally, several tried bringing in objects unrelated to Morandi's work, such as a smartphone, to see how the result would change.

Fig. 7. Example of arrangements by visitors

3.2 Evaluation by Questionnaire

We created an online form where visitors could post their impressions anonymously after their experience, providing the link with a QR code. Approximately one-fifth of the about 600 visitors experienced the exhibit, and 29 people responded. First, we asked if visitor's had prior knowledge of Morandi. 3 respondents (10.3%) said "I am very familiar with him," 5 (17.2%) responded "I have seen his work," 3 (10.3%) selected "I have heard of him," and 18 (62.1%) responded "I did not know anything about him."

Second, we asked the 8 respondents who were most familiar with Morandi "Did you feel the images displayed in the frame as Morandi-like?" In response to this question, four respondents (50%) answered affirmatively, citing the similarity of objects, coloring, and shades to the original.

Third, we asked about changing interest in Morandi. 19 people (65.5%) responded positively. Those who said they did not know Morandi said, "When

I saw his paintings after the experience, I felt a sense of closeness to him." "Until now, I was not very interested in still-life paintings, but I could find the pleasure of imagining the painter's intentions, such as composition and light settings." and "I wanted to know the intention of his work by positioning the object myself.". Those who said they were very familiar with him said, "It led me to wonder what the elements that I thought Morandi were." The three persons who answered negatively (10.3%) replied: "I'm not interested in painting." "There was no explanation on Morandi's art history and status as a painter, so I couldn't understand it." "The exhibit itself was interesting, but my interest in the painter did not increase because it was not my favorite style.".

Free responses indicated that one person voluntarily researched Morandi immediately after viewing the exhibit. Besides this, respondents expressed interest and concern about copyrights, etc., such as "It will lead to fake pictures." "Morandi would not like it. In the future, visual artists should think about rights like AI learning rights." and "Can the images produced by this work be subject to copyright?".

3.3 A Finding at the Exhibition Site

We had a new finding at the exhibition site. As mentioned earlier, CycleGAN is an algorithm that learns by giving a group of images of two domains without pairing them. When an image corresponding to one domain is given, it generates an image corresponding to the other domain. Regarding this exhibit, we gave images of motifs on the table at the exhibition site and images of Morandi's works while training. As a result, if an object, such as a part of a spectator's body or possessions, which was not included in the image group, came into the camera view while the spectator was placing the motifs, the coloring of the generated images was very different from the coloring of Morandi's world. It was as if the magic was broken. This effect is not surprising considering the characteristics of CycleGAN. However, it reminded us of what kind of eyes we have constructed through machine learning. Also, we were able to get the idea that it might be possible to lead to an expression of going back and forth between the world of the author's works and the everyday world.

4 Discussion

Based on these results, we discuss three major points: how effective the exhibit was as a new way to appreciate artwork, the reproduction of features characteristic of Morandi's paintings, and concerns about copyright.

First, regarding the effectiveness of our exhibit as a new way of appreciating Morandi's paintings, over 65% of those who responded to our questionnaire reported increased interest in Morandi's work, with one of them voluntarily researching Morandi after the experience. Moreover, many spectators tried out their own arrangements, 789 variations were made during the exhibition. From these facts, we consider that this exhibit was effective as an opportunity for spectators to become interested in Morandi. Furthermore, in a review of the exhibition that includes this exhibit[7], the chief researcher of a museum that collects several of Morandi's works commented as follows:

> Recently, machine learning has begun to be used for the restoration of artworks. As a person who works in an art museum, I felt that this proposal has potential as an educational tool for understanding artworks and inspiring creativity at the same time.

Based on the results of the questionnaire to spectators and the comments from outside specialists, it appeares that the method proposed in this exhibit is useful as a new way of appreciating Morandi's paintings.

Next, as to whether or not our exhibit was able to reproduce the characteristics of Morandi's painting such as composition, coloring, and contour distortion, half of the respondents who had previously seen Morandi's works said that they felt Morandi's uniqueness, suggesting that a certain level of reproducibility was achieved. Regarding composition, it may be useful to provide support to explore Morandi's unique composition by presenting images of original paintings with similar arrangements. Additionally, allowing the height of the camera and the angle of lights to be easily changed would help the spectators create a more Morandi-like composition. Regarding coloring, our implementation uses simple segmentation. Utilizing segmentation masks created with U-Net, Mask R-CNN, etc. for conversion to an intermediate representation would improve the color quality. Regarding the distortion of the contour, it may be possible to reproduce by correcting intermediate images by referring to the original objects in photographs by Meyerowitz. We want to try out these ideas in the future.

Finally, in terms of copyright, the revised Copyright Act, which came into effect in January 2019 in Japan, added an item which states that it is legal to create machine learning models using images available on the Internet. To make sure, we asked a lawyer who is familiar with AI and data-utilization to review our exhibit, and he told us that it was unlikely to be problematic from a copyright perspective unless it produced images that would be considered identical to copyrighted work. In the first place, our intention was not to create a new work in the style of a deceased author like the Next Rembrandt project, nor is our intention to claim copyright on the images produced. However, as the results of the questionnaire showed concerns about copyright, it is necessary to convey the intent of this exhibit carefully by adding appropriate explanations so as not to be misunderstood.

5 Future Work

In this study, we proposed a new way of appreciating artwork based on its growing interest in the active appreciation and the development of machine

[7] https://www.iamas.ac.jp/report/review-iamas2020/.

learning technology. We address our ideas for future development in the following order, from incremental improvements to higher levels.

First, we convey the issue of the motif trimming. How to trim motifs and background is an essential part of the creative activity of a painter. According to the catalog of Morandi's diverse works, most of his paintings use horizontal canvases but in varying proportions. The motif placement in a canvas also varies; such as placing in the center, on either sides, or so that they protrude from the canvas. Our study did not consider the issue of the trimming directly, but instead we have cut out the center of the collected images as a square. If we increase the number of data channels from 3 channels of red, green, and blue to 4 channels—including alpha transparency—we may grasp characteristics of trimming according to the placement of the motif.

Second, we point up the subject of applying the system to other painters. The french painter Paul Cézanne (1839–1906) left many still-life paintings—giving a significant influence on the painters in the 20th century—where the motifs used in the pictures and the studio used for their production have been preserved as they were. The previously mentioned photographer Meyerowitz also published a book of photographs of Cézanne's motifs and studio [10], describing his commentary as an opportunity to visit Cézanne's studio and to understand why the subject and the background are the same in terms of their flatness. According to Meyerowitz's interpretation, Cézanne painted the walls of the studio in his own tonal dark gray, so the motifs no longer reflect around the edges of the wall nor will it distract distinguishing the objects from the background. As Meyerowitz pointed, intuitively going to the place where the work was created can enhance the understanding of the work than just taking a look at the framed picture. Although recreating an entire studio in a museum exhibition is nearly impossible due to issues with budget or space, it may be possible to recreate them to a sufficient extent by limiting them to a wall and the central motif for an exhibition that aims to encourage spectators to understand the work and provoke creativity.

Third, we see the challenge of archiving not only the final product or the works, but also the process to reach the work. Among the creative acts of the artist, this study dealt with "How would you paint such motifs in front of you?". This values the artist's own perspective from the physical eye, but not the selection or arrangement of motifs. Perhaps if we are able to record all the countless repetitive trial-and-error actions until the final work—such as selecting, placing, evaluating, and starting over, including the decisions made by the artist—we can archive the artist's thoughts on the selection and arrangement of motifs, limited to a particular work.

Finally, machine learning can be an effective method to archive a snapshot of an author of a particular work as an artificial narrow intelligence. First, create an AI that stays close to an author from the stage when the author creates a work. Next, create a dataset by sequentially recording the process leading up to the final work. Then, create a model by learning from the dataset. If such a machine-learning archive could be achieved, it may go beyond just extending the appreciation methods proposed in this study. An example could be on applying to the re-exhibit of "installation" artworks. Installation is a form of artwork and constructed by placing various objects in a particular space. Every time the space in which the work is exhibited changes, the selection and arrangement of objects will also change, which requires the artist to make detailed decisions again. However, if the artist cannot physical be present at site (e.g., has already passed away or is unable to travel to the site due to a global pandemic), setting up the work is a great challenge.

Even when re-exhibiting a work in the same exhibition space, artists encounter difficulties. Our research team involves an artist whose installation works have been exhibited around the world. There was one case where the registrar measured the arrangement of hundreds of objects in millimeters and attempted to re-exhibit the work in the same room where it was first exhibited at a museum that houses his work as a collection, but the registrar could not recreate. Since his work requires the shadows created by lighting objects, the slightest misalignment was magnified. Perhaps an AI that learns the artist's thoughts about a work could make decisions on behalf of the artist. Our research team calls this idea "Archival Archetyping". As we have seen so far, this study is an entry point to many of these promising future developments.

6 Conclusion

In this study, we proposed a new method for the active appreciation of artworks utilizing machine learning. We chose Italian painter Giorgio Morandi as the theme, and developed a hands-on exhibit in which spectators freely arrange objects that reproduce the painter's motifs and compose images with the aid of machine learning that reproduce the coloring of his paintings. From the results of a questionnaire survey of 29 people, we confirmed that experiencing this exhibit deepened their interest in the painter.

References

1. Briot, J.-P., Hadjeres, G., Pachet, F.-D.: Deep learning techniques for music generation - a survey (2017)
2. Deshpande, A., Rock, J., Forsyth, D.: Learning large-scale automatic image colorization. In: Proceedings of the IEEE International Conference on Computer Vision, pp. 567–575 (2015)
3. Elgammal, A.M., Liu, B., Elhoseiny, M., Mazzone, M.: CAN: creative adversarial networks, generating "art" by learning about styles and deviating from style norms. CoRR, abs/1706.07068 (2017). http://arxiv.org/abs/1706.07068
4. Gatys, L.A., Ecker, A.S., Bethge, M.: A neural algorithm of artistic style. CoRR, abs/1508.06576 (2015). http://arxiv.org/abs/1508.06576
5. Gatys, L.A., Ecker, A.S., Bethge, M.: Image style transfer using convolutional neural networks. In: Proceedings of the IEEE Computer Society Conference on Computer Vision and Pattern Recognition, 2016-December, pp. 2414–2423 (2016). ISSN 10636919. https://doi.org/10.1109/CVPR.2016.265

6. Isola, P., Zhu, J.-Y., Zhou, T., Efros, A.A.: Image-to-image translation with conditional adversarial networks. In: Proceedings of the IEEE Conference on Computer Vision and Pattern Recognition, pp. 1125–1134 (2017)
7. Kogan, G.: Artist in the cloud: Towards an autonomous artist (2019)
8. Lu, L.F.L.: A art café: a 3D virtual learning environment for art education. Art Educ. **61**(6), 48–54 (2008)
9. Meyerowitz, J.: Morandi's Objects. Damiani Editore (2015). ISBN 9788862084536
10. Meyerowitz, J.: Cézanne's objects. Damiani Editore (2017). ISBN 9788862085649
11. Morandi, G.: Giorgio Morandi. Silvana Editoriale (2019)
12. Salimbeni, G., Leymarie, F.F., Latham, W.: Generative system to assist the artist in the choice of 3D composition for a still life painting. In: Machine Learning for Creativity and Design (NeurIPS 2019 Workshop) (2019)
13. Di Serio, Á., Ibáñez, M.B., Kloos, C.D.: Impact of an augmented reality system on students' motivation for a visual art course. Comput. Educ. **68**, 586–596 (2013). ISSN 0360-1315. https://doi.org/10.1016/j.compedu.2012.03.002. http://www.sciencedirect.com/science/article/pii/S0360131512000590
14. Smith, G., Leymarie, F.F.: The machine as artist: an introduction. Arts **6**(2), 5 (2017). ISSN 2076-0752. https://doi.org/10.3390/arts6020005. http://www.mdpi.com/2076-0752/6/2/5
15. Turing, A.M.: Computing machinery and intelligence. Mind **59**(236), 433–460 (1950)
16. Yenawine, P.: Visual Thinking Strategies: Using Art to Deepen Learning Across School Disciplines. Harvard Education Press (2013)
17. Zhu, J.-Y., Park, T., Isola, P., Efros, A.A.: Unpaired image-to-image translation using cycle-consistent adversarial networks. In: Proceedings of the IEEE International Conference on Computer Vision, pp. 2223–2232 (2017)

STBM: Stochastic Trading Behavior Model for Financial Markets

Masanori Hirano[1(✉)], Hiroyasu Matsushima[2], Kiyoshi Izumi[1], and Hiroki Sakaji[1]

[1] School of Engineering, The University of Tokyo, 7-3-1, Bunkyo-ku, Hongo, Tokyo 113-8656, Japan
hirano@g.ecc.u-tokyo.ac.jp, {izumi,sakaji}@sys.t.u-tokyo.ac.jp
[2] Center for Data Science Education and Research, Shiga University, 1-1-1, Banba, Hikone, Shiga 522-8522, Japan
hiroyasu-matsushima@biwako.shiga-u.ac.jp
https://mhirano.jp/

Abstract. This is an extension from a selected paper from JSAI2020. In this study, we propose a stochastic model for predicting the behavior of financial market traders. First, using real ordering data that includes traders' information, we cluster the traders and select a recognizable cluster that appears to employ a high-frequency traders' market-making (HFT-MM) strategy. Then, we use an LSTM-based stochastic prediction model to predict the traders' behavior. This model takes the market order book state and a trader's ordering state as input and probabilistically predicts the trader's actions over the next one minute. The results show that our model can outperform both a model that randomly takes actions and a conventional deterministic model. Herein, we only analyze limited trader type but, if our model is implemented to all trader types, this will increase the accuracy of predictions for the entire market.

Keywords: Machine learning · Deep learning · Data mining · High-frequency trade · Market-making · Clustering

1 Introduction

Today, there are increasing systemic risks in the financial market. It is because the complexity of the financial market has been increasing. The financial crisis of 2007–2008 was one of the most famous examples of systemic risks. To avoid future crashes, or more accurately predict future markets, we need to discover the underlying mechanism and take this into account. Many researchers are attempting to predict market movements using machine learning approaches using only macroscopic data, such as market prices and macroeconomic indicators. However, such approaches completely ignore financial markets' micro–macro mechanisms, the most important of which is that movements of the whole market are attributed to the accumulated interactions between individual traders' actions.

K. Yada et al. (Eds.): JSAI 2020, AISC 1357, pp. 157–165, 2021.
https://doi.org/10.1007/978-3-030-73113-7_14

To deal with this problem, we need to build a trader model to predict each trader's actions for simulation. Recently, ordering data with masked trader information from the Tokyo Stock Exchange (TSE) was made available to certain people for particular usages, making it possible to filter the ordering data by traders. Moreover, our proposed clustering method, which extends the approach of [15], enables us to identify traders who show specific trading strategies. In this study, we propose a stochastic model for predicting traders' ordering behavior based on a type of neural network known as the long short-term memory (LSTM). We call this model the Stochastic Trading Behavior Model (STBM). Herein, we test the STBM on data from traders employing the well-known high-frequency trading market-making (HFT-MM) strategy. We focus on this because, as a first step, it is too challenging to handle all possible trader types at once. The HFT-MM strategy involves making frequent limit orders to obtain the margin price between the best ask and the best bid. However, the price is very likely to dramatically change beyond the spread; to hedge against this risk, HFT-MM traders monitor the market and make rapid series of orders, repeatedly making new orders and then canceling them. This strategy is distinctive and relatively easy to be identified in financial market data. So we decided to target HFT-MM traders when testing our STBM.

In our study, the STBM showed promising performance, outperforming both a by-chance model and the theoretical limit on deterministic models at the test for a HFT-MM behavior prediction task. Although these results only focus on HFT-MM traders, if our model can be applied to other types of traders, then it is possible to predict each trader's behavior and make trustable bottom-up simulations or financial market predictions based on real data.

2 Related Work

Over the years, there have been many attempts to predict future markets. Traditionally, technical or fundamental analyses of financial markets were conducted, but owing to technological advances, other prediction approaches are gaining in popularity. First, several researchers have used support vector machines to predict prices or price indices [7,8]. Some studies have also used deep learning to predict market prices, movements, or turning points [3,11,14,17]. Moreover, Wang *et al.* [16] proposed the CLVSA, a model for predicting price trends in financial markets based on the LSTM and Seq2seq deep learning models. Tashiro *et al.* [12] proposed an LSTM-based short-term price prediction model based on ordering data at millisecond timescales. They later extended this method to use a convolutional neural network instead of an LSTM [13]. However, the above work focused only on macroscopic data or mechanisms to predict macro indices such as market trends, price movements, or price indices, ignoring potential micro-mechanisms.

Alternatively, some researchers have taken other approaches, focusing on the traders' actions. Chiarella *et al.* [2] proposed a very basic trader model for use in simulations that determines the trader's strategy based on fundamental factors, trends, and noise. Other researchers later built and demonstrated a model of HFT-MM traders [1] based on equations that consider the market ordering strength, current fundamental prices, and trader inventories. They used this to perform multi-agent financial market simulations.

Moreover, some studies have focused on HFT or HFT-MM strategies. Hosaka [6] performed an HFT analysis based on the same data used in this study, finding that many of the HFTs in the TSE market were executing market-making orders. Uno *et al.* [15] proposed a method of clustering traders based on financial market data (which we extend in this study) and demonstrated that it could identify the orders of traders employing distinctive strategies, such as HFT-MM. Moreover, Hirano *et al.* [4] showed the difference between the behavior of actual HF-MM trader and simulated HFT-MM trader [1] using the same data used in this study.

Other researchers have also studied financial markets using real data. For example, Miyazaki *et al.* [9] detected illegal orders and trades by applying a Gaussian mixture model to raw ordering data, while Nanex [10] reported several distinctive trader behaviors based on mining ordering data.

3 Data and Extracting HFT-MM Orders

In our study, we used two kinds of data on TSE. One is the FLEX Standard[1], which contains the order book information. The other is order book reproduction data[2]. This data contains all complete orders and masked trader information and enables us to trace traders' behaviors. We used ordering data of HFT-MM traders, which is extracted from order book reproduction data. The way of the extraction is explained in the following. The data from January 2015 to July 2015 was used for building our new model. On the other hand, the ordering data from August 2015 was used for evaluation and comparison.

Although order book reproduction data has masked traders' information, it cannot help us to identify the type or strategy of traders. So, to extract ordering data of HFT-MM only, we used a clustering analysis base on [15]. In our paper, we extended the method [15] in terms of the usage of the executed and market order ratio.

At first, for each trader, we calculated the frequency of actions per each stock ticker:

$$(\text{ActionsPerTicker}) = \frac{(\text{newOrders}) + (\text{changeOrders}) + (\text{cancelOrders})}{(\text{numTickers})}. \quad (1)$$

Here, we calculate the index for each business days and we employ the median of the indices of business days in the data. Then to extract HFT only, we filter

1 https://www.jpx.co.jp/english/systems/equities-trading/. This data is available for those who have a contract for the disclosure system.

2 This data is generally not available. It requires a special agreement with Japan Exchange Group.

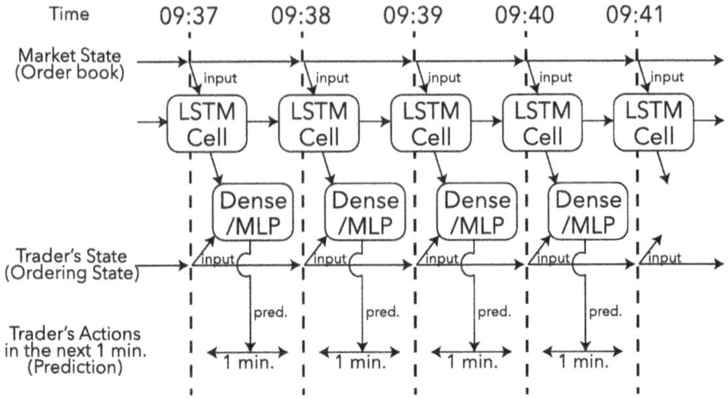

Fig. 1. Overview of our neural network model.

traders by (ActionsPerTicker) > 100. This gave us a total of 181 traders. Then, we also calculated the following indices for each trader on each day.

$$(\text{InventoryRatio}) = \underset{\text{ticker}}{\text{Median}}\left(\left|\frac{(\text{soldVol.})_{\text{ticker}} - (\text{boughtVol.})_{\text{ticker}}}{(\text{soldVol.})_{\text{ticker}} + (\text{boughtVol.})_{\text{ticker}}}\right|\right) \quad (2)$$

$$(\text{ExecutedRatio}) = (\text{executedOrders})/(\text{newOrders}) \quad (3)$$

$$(\text{CancelRatio}) = (\text{cancelOrders})/(\text{newOrders}) \quad (4)$$

$$(\text{MarketOrderRatio}) = (\text{marketOrders})/(\text{newOrders}) \quad (5)$$

$$(\text{TickerPerVSLOG}) = \ln\{(\text{numTikcer})/(\text{numVS})\} \quad (6)$$

$$(\text{ActionsPerTickerLOG}) = \ln(\text{ActionsPerTicker}) \quad (7)$$

After calculating these indices for each day, we took the median values. Then, we normalized all indices over traders before clustering analysis. The clustering analysis we employed was hierarchical clustering with Ward's method, Euclidean distances, and a limit of ten clusters.

As a result, we found one cluster which satisfies all criteria for HFT-MM. In the cluster, there are seven traders. To ensure that these seven traders employ the HFT-MM strategy, we confirmed it by plotting their ordering histories and stock prices. Thus, finally, we got the ordering data of HFT-MM.

4 Trader Behavior Prediction Model

Finally, we built a neural network model to predict the behavior of the traders identified above. This model, based on an LSTM [5], predicts each trader's behavior over the next one minute. We adopted an LSTM to evaluate the market state because it changes over time and the traders' actions depend on historical information, such as trends.

Figure 1 gives an overview of our model. This model is a kind of deep learning model to predict the future behavior of the traders. In this model, we employed an LSTM [5] for processing time-series market states.

As a market state, we input data from FLEX Standard. The orders are divided into 18 classes: (1) best bid, (2–8) 1–7 tick(s) below the best bid, (9) 8 or more ticks below the best bid, (10) best ask, (11–17) 1–7 tick(s) above the best ask, and (18) 8 or more ticks above the best ask. Inputs data are turned into percentages of all volumes on the order book. In addition to these 18 classes, we include two features describing the best bid and ask prices. Thus, LSTM receives a total of 20 input features.

As another input representing the traders' ordering state, we input the detailed order data of HFT-MM. This data contains the orders' placement state at each minute. There are 18 classes in this input, which is the same as the market state. These values are also the percentage of all volume of the current trader's orders.

After the processing by LSTM and Dense/MLP layer in Fig. 1, our model predicts the probabilities of the actions in the next minute. The 21 possible ordering actions are abovementioned 18 classes and (19) above best bid, (20) below best ask, and (21) other (do nothing). For the prediction, we used the softmax layer. And, for the model training, we used Adam optimizer and loss function of mean squared error (MSE).

We trained and validated this model, selecting appropriate hyperparameters, before evaluating its final performance. We divided this data into training, validation, and test sets, using the data from January to July (143 days) for training and validation and the August data (21 days) for testing. In each day, there are some data from the morning session or afternoon session. And in each session, some set of data from different tickers traded by each trader also exist. Each data set from different tickers and traders in the same day and session included the same market state data, but different ordering states and trader actions. This gave us a total of 85,038 data sets for training and validation, and 11,053 data sets for testing. We used 90% of these 85,038 data sets for training and the other 10% for validation. As a result, LSTM with a hidden layer size of 523 with an embedded size of 505, a dense layer with an embedded size of 310, and a ReLU activation function are employed. In this tuning, we also tested multiple layered MLP. However, this result indicates that deep dense/MLP layers are not needed.

After tuning the hyperparameters, we tested our model. Here, we measured the MSE between the predicted actions and ground truth trader actions for the following minute. For example, given data up to 9:38, we made predictions for 9:38 to 9:39 and compared these with the ground truth data for this period. As discussed above, the traders' actions are converted to probabilities, so the predictions are defined similarly, giving us probabilities for the 21 possible trader actions defined above. Particularly, we define the probabilities of taking each of these 21 classes at time t as p_1^t to p_{21}^t, and define the corresponding ground truth values as $\widehat{p_1^t}$ to $\widehat{p_{21}^t}$. These must satisfy $\sum_{i=1}^{21} p_i^t = 1, \sum_{i=1}^{21} \widehat{p_i^t} = 1$, and the MSE at

Table 1. MSE results for three different models.

Model	MSE
STBM	0.1701 ± 0.2750
By-chance	0.3961 ± 0.3622
Ideal deterministic model	0.5301 ± 0.5066

time t is defined as $\mathrm{MSE}_t = \sum_{i=1}^{21} \left(\widehat{p_i^t} - p_i^t\right)^2$. The input data lengths were 150, i.e., one-minute intervals between 9:00 and 11:30 or between 12:30 and 15:00, so we also made 150 predictions for each data set. Below, t represents the number of minutes since the start of the data: the predicted action for, e.g., 9:37–9:38 is the 38th prediction for 9:00–11:30, so $t = 38$. The final MSE is defined as $\mathrm{fMSE} = \frac{1}{150} \sum_{t=1}^{150} \mathrm{MSE}_t$, and is calculated for each data set. (TSE market have two 150 min sessions in one day.) Finally, we employed the mean fMSE value over the entire test data set as the performance index.

For comparison, we considered two other models. The first was a by-chance model; this zero-intelligence model was selected from the 21 classes at random with probability 1.0/21. The second model represented the theoretical limit on deterministic models. We assumed that the model would always take the best possible option, defined as

$$p_i = \begin{cases} 0.5 \text{ (if } i \in \{j, k\}) \\ 0.0 \text{ (other)} \end{cases} \text{where} \begin{cases} j = \arg\max_{i \in \{1, \cdots, 10\}} \{\widehat{p_i}\} \\ k = \arg\max_{i \in \{11, \cdots, 20\}} \{\widehat{p_i}\} \end{cases}. \quad (8)$$

5 Results

Next, we consider the statistical results. Table 1 shows the MSE for each of the three models. Here, our STBM outperforms the other two models, but the significant variances indicate that the prediction accuracies substantially fluctuated. Thus, we also performed t-tests on corresponding samples for each pair of models. t-value between STBM & By-chance, STBM & Deterministic, and By-Chance & Deterministic are 764.480, 777.704, and 474.455, respectively. These results clearly show that our STBM outperformed both the by-chance model and theoretical limit on any deterministic model.

6 Discussion

First, we discuss our STBM's performance, particularly the fact that it outperformed the theoretical limit on any deterministic model. Although we employed the MSE metric where deterministic models have a disadvantage, in real markets, traders cannot determine their actions without resorting to probabilities because future financial markets are inevitably uncertain. Thus, we believe that stochastic action models represent more natural trading strategies. Moreover, our model substantially outperformed the by-chance model.

The advantage of our STBM is that its neural network can account for the market and ordering states. In addition, the LSTM can remember previous market movements when evaluating the market state, enabling it to make more accurate and detailed predictions than conventional models. However, such data-driven approaches have only recently become possible, owing to the huge volume of data required. Moreover, as previously mentioned, such trader-based predictions have only become possible owing to the availability of new data that includes masked trader information.

This work can also be extended to improve conventional financial simulations, which depend heavily on manually created trader models which may, therefore, be biased. Moreover, many simulation-based approaches are deterministic. With our STBM, financial simulations will become more trustable and realistic.

In future work, we plan to improve our model in three main areas. First, we also plan to test other neural network architectures. We could consider other possible architectures, such as a convolutional neural network, instead of a dense layer/MLP.

Second, we should apply the STBM to other types of traders. Here, we focused on HFT-MM traders, selected via a clustering analysis. However, if we are to predict the movements of the whole market by combining the predicted actions of all traders, we will have to consider other trader types; this will imply overcoming several obstacles. For example, currently, the STBM only predicts the action probabilities for the next minute, but we also need to consider the number of orders made during that time.

Third, to combine our STBM with simulation-based approaches for financial markets, we should conduct tests to confirm that it also correctly works under simulated conditions, as real and simulated financial markets may be somewhat different. However, if we can combine an STBM based on real market data with hypothetical market simulations, we can potentially make predictions for scenarios that have never occurred.

7 Conclusion

In this study, we have proposed the STBM for financial markets. This study aims to predict a trader's actions during the next minute based on the previous market state and the trader's ordering state. It is a neural network model based on an LSTM and predicts the trader's actions probabilistically with the possibility of doing nothing. Here, we focused on applying the STBM to traders employing high-frequency trading market-making (HFT-MM) strategies. First, using real ordering data that included masked trader information (order-book reproduction data), we performed a hierarchical clustering analysis only for high-frequency traders, selecting one cluster that we believed to consist of HFT-MM traders. Then, we used the data from these traders to train an STBM. The results show that it outperformed both a by-chance model and the theoretical limit on conventional deterministic models. In future work, we plan to apply the STBM to other trader types. In addition, we could apply it to agent-based simulations.

Acknowledgement. We thank the Japan Exchange Group, Inc. for providing the data.This research was supported by MEXT via Exploratory Challenges on Post-K computer (study on multilayered multiscale space-time simulations for social and economic phenomena) and the computational resources of the HPCI system provided by the Information Technology Center at The University of Tokyo, and the Joint Center for Advanced High Performance Computing through the HPCI System Research Project (Project ID: hp190150).

References

1. Avellaneda, M., Stoikov, S.: High-frequency trading in a limit order book. Quant. Financ. **8**(3), 217–224 (2008)
2. Chiarella, C., Iori, G.: A simulation analysis of the microstructure of double auction markets. Quant. Financ. **2**(5), 346–353 (2002)
3. Dixon, M.F., Polson, N.G., Sokolov, V.O.: Deep learning for spatio temporal modeling: dynamic traffic flows and high frequency trading. Quant. Financ. **19**(4), 549–570 (2019)
4. Hirano, M., Izumi, K., Matsushima, H., Sakaji, H.: Comparing actual and simulated HFT traders' behavior for agent design. J. Artif. Soc. Soc. Simul. **23**(3), 6 (2020). https://doi.org/10.18564/jasss.4304
5. Hochreiter, S., Schmidhuber, J.: Long short-term memory. Neural Comput. **9**(8), 1735–1780 (1997)
6. Hosaka, G.: Analysis of high-frequency trading at Tokyo stock exchange. In: JPX Working Paper (2014)
7. Kercheval, A.N., Zhang, Y.: Modelling high-frequency limit order book dynamics with support vector machines. Quant. Financ. **15**(8), 1315–1329 (2015)
8. Kim, K.: Financial time series forecasting using support vector machines. Neurocomputing **55**, 307–319 (2003)
9. Miyazaki, B., Izumi, K., Toriumi, F., Takahashi, R.: Change detection of orders in stock markets using a Gaussian mixture model. Intell. Syst. Acc. Financ. Manage. **21**(3), 169–191 (2014)
10. Nanex: Nanex - Market Crop Circle of The Day (2010). http://www.nanex.net/FlashCrash/CCircleDay.html
11. Sirignano, J.A.: Deep learning for limit order books. Quant. Financ. **19**(4), 549–570 (2019)
12. Tashiro, D., Izumi, K.: Estimating stock orders using deep learning and high frequency data [in Japanese]. In: Proceedings of the 31nd Annual Conference of the Japanese Society for Artificial, p. 2D2-2 (2017)
13. Tashiro, D., Matsushima, H., Izumi, K., Sakaji, H.: Encoding of high-frequency order information and prediction of short-term stock price by deep learning. Quant. Financ. **19**(9), 1499–1506 (2019)
14. Tsantekidis, A., Passalis, N., Tefas, A., Kanniainen, J., Gabbouj, M., Iosifidis, A.: Using Deep learning to detect price change indications in financial markets. In: Proceedings of 25th European Signal Processing Conference, pp. 2580–2584 (2017)

15. Uno, J., Goshima, K., Tobe, R.: Cluster analysis of trading behavior: an attempt to extract HFT [in Japanese]. In: The 12th Annual Conference of Japanese Association of Behavioral Economics and Finance (2018)
16. Wang, J., Sun, T., Liu, B., Cao, Y., Zhu, H.: CLVSA: a convolutional LSTM based variational sequence-to-sequence model with attention for predicting trends of financial markets. In: Proceedings of the Twenty-Eighth International Joint Conference on Artificial Intelligence (IJCAI-19), pp. 3705–3711 (2019)
17. Zhang, L., Aggarwal, C., Qi, G.J.: Stock price prediction via discovering multi-frequency trading patterns. In: Proceedings of the 23rd ACM SIGKDD, pp. 2141–2149 (2017)

Gated Extra Memory Recurrent Unit for Learning Video Representations

Daria Vazhenina and Atsunori Kanemura(✉)

LeapMind Inc., Tokyo 150-0044, Japan
{daria,atsu-kan}@leapmind.io

Abstract. Convolutional recurrent neural networks (ConvRNNs) are widely used for spatiotemporal modeling tasks including video frame prediction. A major drawback of existing ConvRNNs is the amount of computing and memory resources, which can hinder practical applications on embedded devices. Thus, to reduce them, we propose 1) a new gated architecture of the recurrent unit with temporal memory and 2) the replacement of computationally demanding convolution with a more light-weight Hadamard product. Adopting such constraints can degrade the performance, but we show that the proposed model produces better results with reduced computation and memory. Quantitative evaluation with the Moving MNIST dataset shows that the overall performance of video frame prediction is improved by 13% in terms of MSE and by 3% in terms of SSIM without increasing the number of parameters and their multiplications, compared with the conventional ConvLSTM baseline. Further, applying the Hadamard product replacement outperforms the baseline MSE by 5%, while reducing the number of parameters by 14% and the number of multiplications by 25%.

1 Introduction

Learning video representations allows us to model dynamical structures by observing events in videos. A model that can predict future observations is likely to have acquired a useful representation for a variety of visual perception tasks, such as object tracking and action recognition [2,6]. However, the computational load of such models creates an obstacle to putting them on embedded devices and applying them to real-world applications.

Video frames (image sequences) contain both spatial and temporal information, which is commonly modeled by a variant of convolutional recurrent neural networks (ConvRNNs). The most common baseline model is ConvLSTM, where three gates are used to support temporal information flow. Although fewer gated versions of conventional RNNs provide superior performance on specific tasks [4], their convolutional counterparts did not provide a similar advantage in video processing tasks [1]. These models tend to use very short context for the next frame prediction, thus keeping shapes after object occlusion becomes a big source of errors. Therefore, increasing the number of gates and making memory units more complicated is the latest trend [8–11].

K. Yada et al. (Eds.): JSAI 2020, AISC 1357, pp. 166–174, 2021.
https://doi.org/10.1007/978-3-030-73113-7_15

In [10], it was proposed to use additional memory to create additional information flow between layers, which requires additional gates. Although it improved overall performance, the computational load was twice as high as that of ConvLSTM. This idea was further developed in [11], which improved information flow by adding more connections to newly created memories, increasing the number of parameters to three times that of a ConvLSTM baseline. As would be expected, an increased number of parameters provides more accurate predictions. But this success comes with a price: high memory requirements and increased computational complexity.

In this paper, we propose a ConvRNN unit with additional memory and a reduced number of gates. Gated extra memory recurrent unit (GEM-RU) has two memory units to keep temporal changes within frame sequence, and two gates to support the information flow. The resulting unit allows us to achieve 3% higher performance in terms of SSIM and 13% lower in terms of MSE compared to ConvLSTM on the Moving MNIST dataset. Further, applying the Hadamard product replacement, as was proposed in [7], outperforms the baseline MSE by 5%, while reducing the number of parameters by 14% and the number of multiplications by 25% compared to ConvLSTM baseline. We performed all experiments using an encoder-decoder model setup, where each part is a stack of CNN and ConvRNN layers.

We also compared GEM-RU and its variants performance with the state-of-the-art unit PredRNN++ [8]. Although, GEM-RU yields performance gains by 3% in terms of SSIM, it has 40% fewer numbers of parameters and multiplications compared with PredRNN++.

2 Related Work

Although a lot of work has been done to improve ConvRNNs performance, ConvLSTM is still the most commonly used baseline unit [5]. The main difference between conventional and convolutional LSTM units is replacing matrix multiplication with convolution operation and changing the shape of input X_t from 1D to 2D. Gating architecture allows us to control information flow within the ConvLSTM unit and can be described by the following equations:

$$i_t = \sigma(X_t * W_{xi} + h_{t-1} * W_{hi} + b_i), \tag{1}$$

$$f_t = \sigma(X_t * W_{xf} + h_{t-1} * W_{hf} + b_f), \tag{2}$$

$$o_t = \sigma(X_t * W_{xo} + h_{t-1} * W_{h0} + b_o), \tag{3}$$

$$c_t = f_t \odot c_{t-1} + i_t \odot \tanh(X_t * W_{xc} + h_{t-1} * W_{hc} + b_c), \tag{4}$$

$$h_t = \tanh(c_t \odot o_t). \tag{5}$$

Here $\sigma(\cdot)$ is the sigmoid function, $*$ means convolutional operation, and $\odot$ is the Hadamard (element-wise) product; $W_{\circ\circ}$ is a set of convolutional kernels. h_t and h_{t-1} are current and previous hidden states, respectively, and c_t and c_{t-1} are current and previous cell outputs, respectively; i_t, f_t, and o_t are input, forget, and output gates, respectively, all of which have the same size as the hidden states. As the values of the gates lie in the $(0, 1)$ interval, they allow the memory units c_t and h_t to store and access information over many training examples.

One of the major limitations with ConvLSTM is keeping object shapes after occlusion by other objects or when the object reappears in the video frame skipping some of them [8]. After these kinds of events, predicted images become blurry for all objects or lose track of one or more objects in a predicted frame.

Recently, it was argued that increasing the number of memory units results in improved ability to keep the shape of all objects in a frame sequence [8–11]. It was proposed to add a memory unit to pass information between layers while keeping the two memory units (cell and hidden state) that pass information over time the same as in ConvLSTM [8,10]. This resulted in performance improvement in terms of MSE and SSIM. However, to improve information flow for inter-layer memory unit, three additional gates were introduced similar to i_t, f_t, and o_t in (1)–(3), which resulted in significant increase of computational and memory load.

For the conventional gated RNN units, it was proposed to reduce the number of gates in the LSTM unit and use only one memory unit over time, like in GRU. This reduced computational load and improved performance for some tasks. Similar to conventional RNN units, the performance of fewer gated convolutional RNN units was investigated [1,3]. Although it provided the reduction of the expected computational load, the issue of keeping object shapes after occlusion still exists in those units.

3 Gated Extra Memory Recurrent Unit

Ideally, a predictive model should be able to learn relevant structures within a video frame and be small and fast enough for putting on an embedded device and applying to a real-world application. Increasing the number of memory units and gates enables a predictive model to learn more disentangled representations and predict both correct shapes and trajectories of moving objects [8]. Reducing the number of gates allows reducing the number of parameters and multiplications while keeping sufficient information flow to avoid the vanishing gradient problem [3]. A natural extension of these ideas is increasing the number of memory units while reducing the number of gates to one per memory unit. We name the resulting unit *gated recurrent unit with extra memory (GEM-RU)*, which is described with the following formulae:

$$r_t = \sigma(X_t * W_{xr} + h_{t-1} * W_{hr} + c_{t-1} * W_{cr} + b_r), \tag{6}$$

$$u_t = \sigma(X_t * W_{xu} + h_{t-1} * W_{hu} + c_{t-1} * W_{cu} + b_u), \tag{7}$$

$$c_t = \tanh(X_t * W_{xc} + (u_t \odot c_{t-1}) * W_{cc} + b_c), \tag{8}$$

$$h_t = r_t \odot \tanh(X_t * W_{xh} + h_{t-1} * W_{hh} + (c_t - c_{t-1}) * W_{hc} + b_h). \tag{9}$$

Here we use similar notation for inputs, memory units, and weight kernel sets as in Sect. 2; r_t and u_t are gates to support information flow for cell c_t and hidden state h_t. Both previous states of memory units are used to calculate gates, as

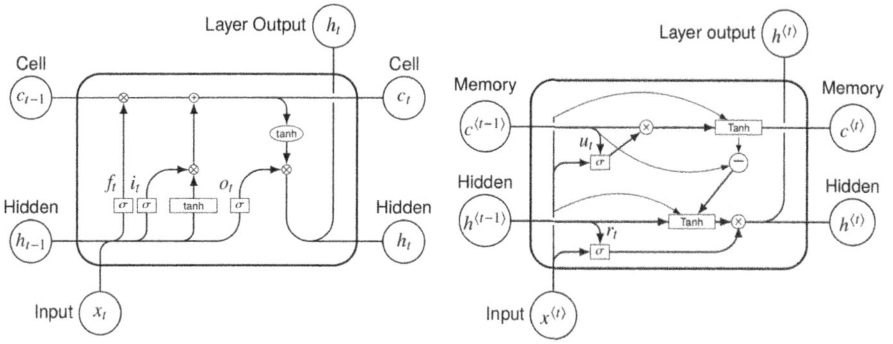

Fig. 1. Left: ConvLSTM recurrent unit architecture. Right: GEM-RU architecture.

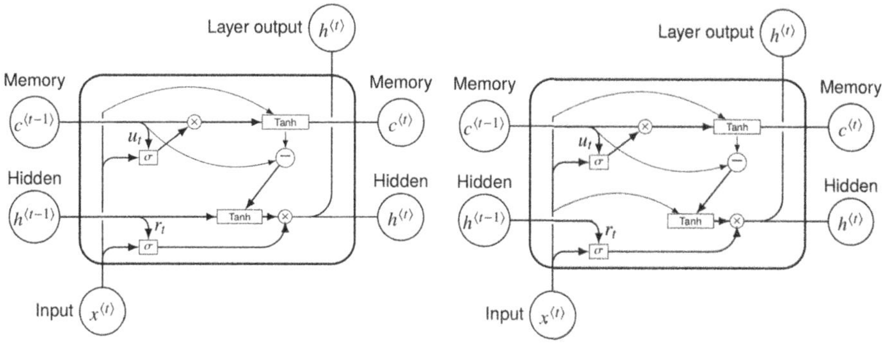

Fig. 2. Left: GEM-RU-h architecture. Right: GEM-RU-i architecture.

memory units in GEM-RU are more independent compared to ConvLSTM. To force the hidden state to track changes within cell memory, we subtract the current cell state from the previous one $(c_t - c_{t-1})$ in the third term of (9).

Although GEM-RU uses fewer gates than ConvLSTM, it has a larger number of convolutional operations, which results in a higher number of parameters. As depicted in Fig. 1, GEM-RU architecture has more complex connections between memory units c_t and h_t, which is implemented to combat the problem of keeping shapes of all objects in a frame.

3.1 Proposed Unit

To reduce the computational load of the GEM-RU we investigate possibilities to reduce the number of convolutional operations. First, we isolate exposure of each gate r_t and u_t to the respective previous state of the memory unit, where those can be described with the following formulae:

$$r_t = \sigma(X_t * W_{xr} + h_{t-1} * W_{hr} + b_r), \tag{10}$$

$$u_t = \sigma(X_t * W_{xu} + c_{t-1} * W_{cu} + b_u). \tag{11}$$

However, the final hidden state h_t in (9) uses three sets of convolutional kernels $W_{\circ\circ}$, while other members of GEM-RU unit uses two sets of kernels $W_{\circ\circ}$ as described by (8), (10) and (11). Second, we investigate two variants of final hidden state h_t: 1) we drop term $X_t * W_{xh}$ in (9) and refer to this unit as GEM-RU-h, 2) we drop term $h_{t-1} * W_{hh}$ in (9) and refer to this unit as GEM-RU-i. As shown in Fig. 2, GEM-RU-h and GEM-RU-i differ only in one connection when calculating memory unit h_t, which is used as input to the next layer and as previous temporal state. These two proposed units have the same number of parameters and multiplications as the baseline ConvLSTM unit.

3.2 IndRNN Extension

Another way to reduce the computational load of ConvRNNs is by substituting some convolution operations with the Hadamard product. It was investigated in [7] that substituting the second convolution in all gate computations allows reducing computational load with a minor performance drop. This combined ConvIndConvLSTM unit provided better performance, we use it as a baseline and refer to it as CIC-LSTM. We apply the same approach for our proposed units, and substitute the second convolutional operation in (10) and (11) with the Hadamard product, resulting in CIC-GEM-RU-h and CIC-GEM-RU-i units. Calculations of memory units c_t and h_t are kept the same.

4 Experimental Setup and Results

4.1 Database Description

We used the moving MNIST dataset [6], which has been widely used to evaluate video frame prediction models. We used the same training and test settings as in [6]. Each sequence in this dataset consists of 20 image frames of size 64×64 with two random moving digits from the MNIST dataset. The performance of the next frame prediction was measured by MSE (the lower this value the better) and SSIM (the higher this value the better) between prediction and the ground truth [12].

4.2 Model Description

To evaluate the proposed units, we used the same composite model as in [7], which has an encoder-decoder pipeline. In the encoder part, we stack two recurrent layers on the top of two CNN layers and revert those recurrent and CNN layers in the decoder part. In all the recurrent layers, the kernel size was 3×3 and the number of channels is 64.

4.3 Performance Evaluation

First, we evaluated our proposed units in terms of MSE and SSIM. Figure 3 shows that our proposed units achieve nearly state-of-the-art performance in

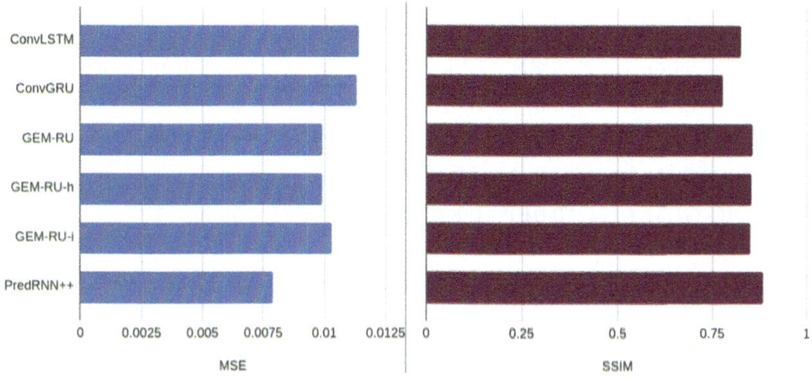

Fig. 3. GEM-RU model performance evaluation in terms of MSE and SSIM. ConvLSTM [5] is included here as a baseline model and PredRNN++ [8] as a state-of-the-art model.

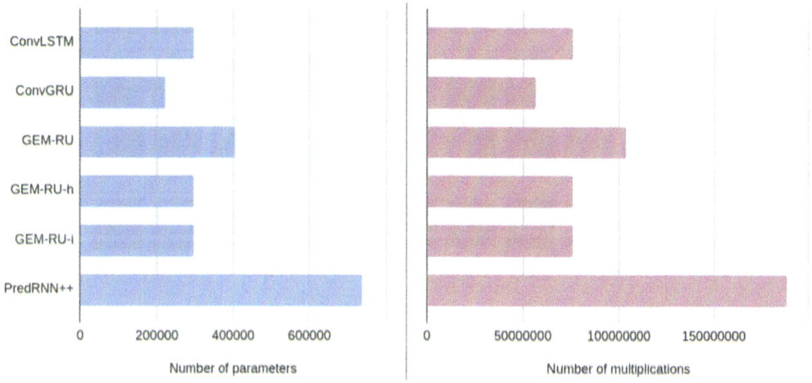

Fig. 4. GEM-RU model performance evaluation in terms of number of parameters and multiplications. ConvLSTM [5] is included here as a baseline model and PredRNN++ [8] as a state-of-the-art model.

terms of SSIM while keeping the number of parameters and multiplications the same as in the baseline, as shown in Fig. 4. GEM-RU and GEM-RU-h have almost equal performance both in terms of MSE and SSIM, but the numbers of parameters and multiplications are fewer in the GEM-RU-h unit. This shows that GEM-RU is over-parameterized compared to GEM-RU-h. Therefore, careful selection of information flows described in Sect. 3.1 allows similar performance to be achieved while using fewer parameters.

In the second experiment, we evaluated CIC (ConvIndConv) variants of our proposed units. The results in Table 1 demonstrate that our proposed units with a reduced number of parameters achieve superior results than the two baseline units, ConvLSTM and CIC-LSTM. As was presented in [7], Hadamard product replacement results in some drop in performance while reducing the number

Table 1. Performance evaluation of the CIC-* models.

Unit name	MSE	SSIM	#Parameters	#Multiplications
ConvLSTM	0.0114	0.8238	294 912	75 497 472
CIC-LSTM	0.0119	0.8111	233 472	47 235 072
CIC-GEM-RU-h	0.0104	0.8281	253 952	56 665 872
CIC-GEM-RU-i	0.0108	0.8258	253 952	56 665 872

Table 2. Video prediction examples with entangled digits in the last two input frames and the first four output frames.

Model name	Example output (frames 10-20)	Number of parameters
Ground truth		
ConvLSTM	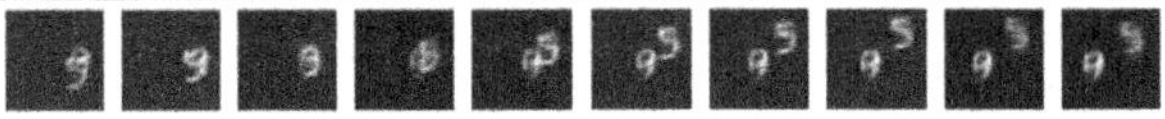	294 912
PredRNN++ [8]		737 280
GEM-RU-h		294 912
CIC-GEM-RU-h		253 952

of parameters. Although the performance of the CIC variant of the proposed units is decreased compared to fully convolutional counterparts, it is superior to the fully convolutional baseline ConvLSTM. CIC-GEM-RUs outperformed the baseline MSE by 5% while reducing the number of parameters by 14% and the number of multiplications by 25%.

Finally, we perform a qualitative analysis of the proposed units. To demonstrate that GEM-RU variants are capable of keeping objects' shapes, we selected a sequence from the test set such that the middle frames contain overlapping digits. Table 2 summarizes example outputs and the number of parameters per ConvRNN layer. Both GEM-RU-h and CIC-GEM-RU-h models predict future frames from entangled digits keeping shapes of both digits better than the ConvLSTM model.

Both quantitative and qualitative analyses demonstrate that GEM-RU-h, GEM-RU-i, and all CIC-GEM variants outperform the baseline ConvLSTM model. The number of parameters was at maximum the same as in the baseline model. This allows us to suggest that improvement comes due to the proposed new architecture.

5 Conclusion

In this paper, we proposed a new gated ConvRNN unit, which allowed us to overcome the issue of keeping object shape. This was achieved by careful design of information flow between gates and memory units. We demonstrated that the proposed GEM-RU variants increase overall model performance while reducing computational and memory load. Further experiments on reducing parameters by replacing some convolutions with the Hadamard product showed that the proposed units are advantageous compared to a ConvLSTM baseline.

Although GEM-RUs yield performance gains compared to the state-of-the-art PredRNN++ unit [8], GEM-RU has about 40% fewer parameters and multiplications. GEM-RU-h variant has about the same performance as GEM-RU, while its computational load is 60% smaller than that of PredRNN++. This suggests that the current trend of increasing the number of memory units is severely over-parameterized and computational load can be optimized.

References

1. Ballas, N., Yao, L., Pal, C., Courville, A.: Delving deeper into convolutional networks for learning video representations. Presented at the (2016)
2. Denton, E., Birodkar, V.: Unsupervised learning of disentangled representations from video. In: Advances in Neural Information Processing Systems (NIPS), pp. 4414–4423 (2017)
3. Elsayed, N., Maida, A.S., Bayoumi, M.: Reduced-gate convolutional LSTM using predictive coding for spatiotemporal prediction. arXiv:1810.07251 (2018)
4. Ravanelli, M., Brakel, P., Omologo, M., Bengio, Y.: Light gated recurrent units for speech recognition. IEEE Trans. Emerg. Topics Comput. Intell. **2**(2), 92–102 (2018)
5. Shi, X., Chen, Z., Wang, H., Yeung, D.-Y., Wong, W., Woo, W.: Convolutional LSTM network: a machine learning approach for precipitation nowcasting. In: Advances in Neural Information Processing Systems (NIPS), pp. 802–810 (2015)
6. Srivastava, N., Mansimov, E., Salakhudinov, R.: Unsupervised learning of video representations using LSTMs. In: International Conference Machine Learning (ICML), pp. 843–852 (2015)
7. Vazhenina, D., Kanemura, A.: Reducing the number of multiplications in convolutional recurrent neural networks (ConvRNNs). In: Ohsawa, Y. (ed.) JSAI 2019. AISC, vol. 1128, pp. 45–52. Springer, Cham (2020). https://doi.org/10.1007/978-3-030-39878-1_5
8. Wang, Y., Gao, Z., Long, M., Wang, J., Yu, P.S.: PredRNN++: towards a resolution of the deep-in-time dilemma in spatiotemporal predictive learning. Presented at the (2018)
9. Wang, Y., Jiang, L., Yang, M.H., Li, L.J., Long, M., Fei-Fei, L.: Eidetic 3d lstm: a model for video prediction and beyond. Presented at the (2019)
10. Wang, Y., Long, M., Wang, J., Gao, Z., Yu., P.S.: PredRNN: recurrent neural networks for predictive learning using spatiotemporal LSTMs. In: Advances in Neural Information Processing Systems, (NIPS), pp. 879–888 (2017)

11. Wang, Y., Zhang, J., Zhu, H., Long, M., Wang, J., Yu, P.S.: Memory in memory: a predictive neural network for learning higher-order non-stationarity from spatiotemporal dynamics. In: IEEE Conference on Computer Vision and Pattern Recognition (CVPR), pp. 9154–9162 (2019)
12. Wang, Z., Bovik, A.C., Sheikh, H.R., Simoncelli, E.P.: Image quality assessment: from error visibility to structural similarity. IEEE Trans. Image Process. **13**(4), 600–612 (2004)

Balancing Policy Improvement and Evaluation in Risk-Sensitive Satisficing Algorithm

Hiroaki Wakabayashi, Takumi Kamiya, and Tatsuji Takahashi(✉)

Tokyo Denki University, Ishizaka, Hatoyama, Hiki, Saitama 350-0394, Japan
tatsujit@mail.dendai.ac.jp
https://sites.google.com/site/tatsujitakahashi/

Abstract. Reducing the search space is one of the challenges in reinforcement learning. One of the satisficing reinforcement learning algorithms, commonly known as RS+GRC, reduces large search space by setting an aspiration level. However, a lag between policy evaluation and improvement, due to policy feedback, prevents proper exploration. Therefore, we propose an eligibility trace-based RS(λ) method, which eliminated the lag. We demonstrated that RS(λ) exhibited efficient learning toward behavior policy-based satisfaction.

1 Introduction

The reinforcement learning aims to maximize the returns. However, in an unknown environment, deciding on knowledge-based actions is necessary while exploring information in the environment. The tradeoff between exploration and exploitation is one of the principal challenges in reinforcement learning. Moreover, optimal action choices are required when aiming to maximize returns. However, determining the unknown optimal choices requires complete exploration of the environment, which may be impractical for most of the tasks.

Human intervention can alleviate learning difficulty by changing the target from the optimal solution to a satisficing solution and by setting up a target aspiration level for such a large search space [3]. Learning by risk-sensitive satisficing (RS) [4,5] and global reference conversion (GRC) [2,6] automatically adjusts the exploitation and exploration, according to the behavior policy returns, by setting task-specific aspiration level. However, the lag between policy evaluation and behavior policy improvement in the current RS+GRC renders the searching difficult.

In this study, we propose RS(λ) method that employed the RS+GRC's eligibility trace and eliminated the said lag. We demonstrated a learning efficiency improvement in a sub-optimal goal environment and identified the factors that contributed to this improvement.

K. Yada et al. (Eds.): JSAI 2020, AISC 1357, pp. 175–182, 2021.
https://doi.org/10.1007/978-3-030-73113-7_16

2 Satisficing Reinforcement Learning

Q-learning method represented reinforcement learning. In Q-learning, for a pair of states s and actions a at a time t, a value was learned using Eq. 2, for selecting appropriate action.

$$\delta^{(t)} = r_{t+1} + \gamma \max_a Q(s_{t+1}, a) - Q(s_t, a_t) \tag{1}$$

$$Q(s_t, a_t) \leftarrow Q(s_t, a_t) + \alpha\delta^{(t)}, \tag{2}$$

wherein α and γ represent the learning and discount rates, respectively.

Target policy maximized the value, while *behavior policy* selected the actions to gather environment-specific information. Acquiring sufficient knowledge of the environment through searching is required to obtain the optimal solutions. The extent of searching is also important to obtain optimal solution. This was a major limitation for reinforcement learning.

In the RS+GRC, the environment-specific global aspiration $\aleph_G$ (aleph) and the difference between the aspiration and the actual profit were employed to obtain a satisfiable policy that exceeded $\aleph_G$.

2.1 Risk-Sensitive Satisficing (RS)

RS learned to obtain the target policy returns that were in agreement with $\aleph_G$. Satisfaction was determined by moving the behavior policy closer to the target policy. The distance between the behavioral and target policies was controlled using GRC method. The GRC method enhanced the local aspiration function, $\aleph_L$, which was the difference between $\aleph_G$ and the of the target policy returns, E_G. It reduced the extent of search when closer to 0 and increased with the distance between the aspiration level and the target policy, which was calculated using the formula 4.

$$\delta_G = \min(E_G - \aleph_G, 0) \tag{3}$$

$$\aleph_L(s_i) = \max_a Q(s_i, a) - \delta_G \tag{4}$$

As we aimed to obtain a satisfactory solution, $\delta_G > 0$ was not considered, because it altered the behavior policy to emphasize $\max_a Q(s_i, a)$.

The behavior policy π in RS was expressed using $\aleph_L$, in Eq. 5.

$$\pi(s, \aleph_L) = \arg\max_a RS(s, a, \aleph_L) \tag{5}$$

$$RS(s, a, \aleph_L) = \tau(s, a)\delta_L(s, a, \aleph_L) \tag{6}$$

$$\delta_L(s, a, \aleph_L) = Q(s, a) - \aleph_L(s) \tag{7}$$

RS qualitative valuation varied based on the sign on aspiration and Q value differences. It considered the correlation between the Q value and τ, which approximated the number of times an action was selected. τ was defined as follows:

$$\tau(s_t, a_t) = \tau_{\text{curr}}(s_t, a_t) + \tau_{\text{post}}(s_t, a_t) \tag{8}$$

$$\tau_{\text{curr}}(s_t, a_t) \leftarrow \tau_{\text{curr}}(s_t, a_t) + 1 \tag{9}$$

$$\begin{aligned} tau_{\text{post}}(s_t, a_t) \leftarrow & (1 - \alpha_\tau)\tau_{\text{post}}(s_t, a_t) \\ & + \alpha_\tau \gamma_\tau \tau(s_{t+1}, \pi(s_{t+1}, \aleph(s_{t+1}))), \end{aligned} \tag{10}$$

wherein γ_τ and α_τ represented the discount and learning rates, respectively.

3 Issues with RS and GRC

The RS+GRC evaluated a behavior policy, using the satisfaction level δ_G and improved it using $\aleph_L$. A gap existed in the evaluation and improvement of the behavior policy for next-state value-utilizing asymptotic value updates (Eq. 1).

The π was altered in the subsequent episodes in an exploiting or exploring manner, following $\aleph_L$, which was updated by the aspiration level δ_G. This variation implied proper policy improvement in previous episodes.

$$E_G \leftarrow \frac{E_{\text{tmp}} + \gamma_G N_G E_G}{1 + \gamma_G N_G} \tag{11}$$

$$N_G \leftarrow 1 + \gamma_G N_G \tag{12}$$

4 Proposed Solutions for the Problem

To regulate the searching practices, we employed RS(λ), which introduced eligibility traces for value enhancement. Eligibility trace [1] calculated the eligibility weights and updated them with trajectory TD errors.

The Watkins's eligibility $e_{s,a}^{(t)}$ was defined as follows:

$$e_{s,a}^{(t)} = \begin{cases} 0, & \text{if } Q_{t-1}(s_t, a_t) \neq max_a Q_{t-1}(s_t, a) \\ \gamma\lambda e_{s,a}^{(t-1)} + 1, & \text{if } s = s_t, a = a_t \\ \gamma\lambda e_{s,a}^{(t-1)}, & \text{otherwise.} \end{cases} \tag{13}$$

Here, λ is a parameter that dampens the amount of updates the more past experience. The action values were calculated using the eligibility $e_{s,a}^{(t)}$ and the TD error δ for all states s and actions a as follows:

$$Q_{s,a} \leftarrow Q_{s,a} + \alpha\delta_{s,a}^{(t)} e_{s,a}^{(t)}. \tag{14}$$

Q learning method updated the values using Eq. 2. Q(λ) updated the following Eq. 14. Watkins's Q(λ)did not trace prior exploratory actions during update and was not a disadvantage for RS. RS estimated the values accurately similar to $maxQ$ with a confidence level of τ and was optimistic about the subsequent episodes. The RS that adopted Watkins's Q(λ) for the value updates were known as RS(λ).

5 Experiment

Here, we used SuboptimaWorld, which had multiple sub-optimal solutions (Fig. 1). The dark gray denotes the start and the light gray denotes the start position for the agent. The agent aimed to obtain the maximum reward of 8; the reward was between 1 and 8. The agent moved to the next square by moving either up, down, left, or right in a single action. The SuboptimaWorld was employed owing to its several sub-optimal goals. Here, the agents were directed to learn an optimal goal path that could evade sub-optimal solutions.

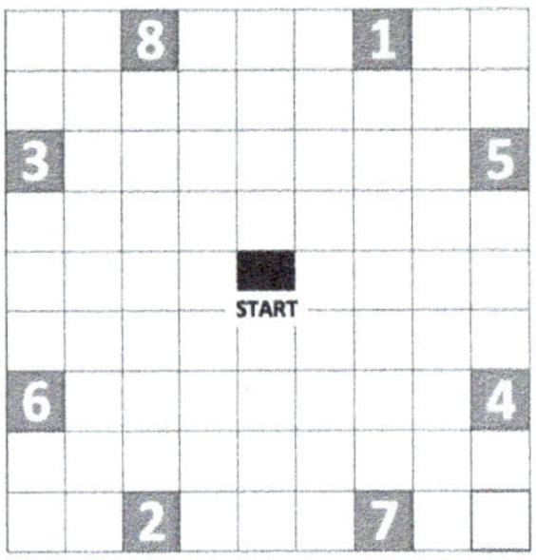

Fig. 1. SuboptimaWorld task

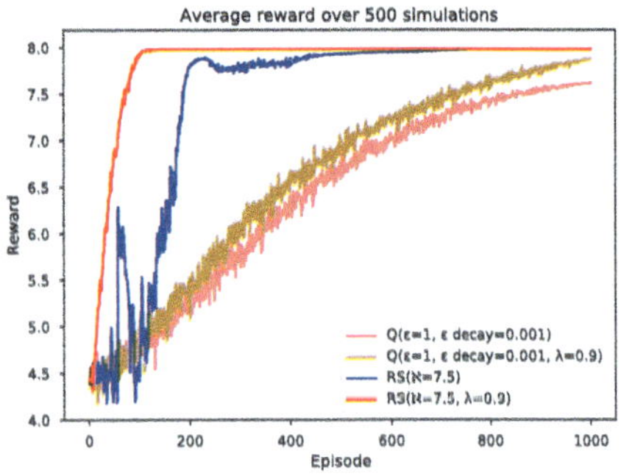

Fig. 2. The performance of the algorithms

5.1 Simulation

Q-learning, Q-learning with eligibility trace, RS+GRC, and RS(λ) methods were used for comparison. Q-learning was operated with ϵ-greedy policy, in which the agent selected a random action at probability ϵ and greedy action a having the highest action value $Q(s_t, a_t)$ at probability $1-\epsilon$. ϵ initiated at 1.0, decreased by 0.001 per episode, and reached 0 at episode 1000. The parameters $\aleph_G = 7.5$ and $\gamma_G = 0.9$ were used for RS+GRC. The parameter, λ, used for eligibility trace, was fixed at 0.9. The learning rate α in all the methods was 0.1. The discount rate γ for all the algorithms was set to 0.9. Each episode ended when the agent either reached one of the eight goals or performed 1000 actions. The experiment was conducted for 1000 episodes and 500 simulations.

5.2 Results

Figure 2 depicts the performance differences between regular RS and the RS(λ). While normal RS exhibited fluctuating earned rewards' average till 200 episodes, the RS(λ) identified the highest averaged reward at 150th episode.

6 Discussion

RS(λ) search exhibited improved learning efficiency than that by RS.

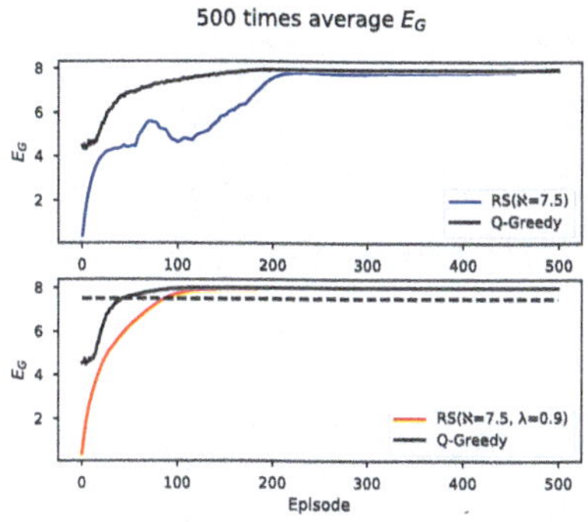

Fig. 3. Gap between the ideal and the present situations. The upper and lower parts depict the normal RS and RS(λ), respectively. The solid black and blue lines depict the target policy observations and agent's actual behavior, respectively.

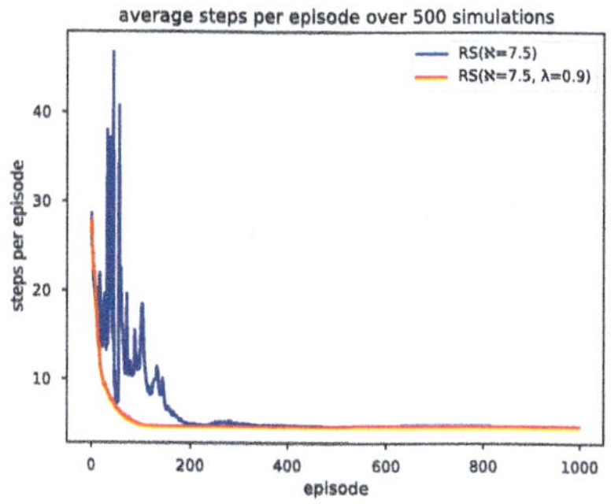

Fig. 4. Number of steps taken per episode.

Evaluation policy is illustrated in Fig. 3. E_G remained at about 4.5 in the normal RS training. This was the average reward for each goal. However, RS(λ) was unstable, and E_G increased with the target policy. The RS and RS(λ) values expanded slowly during the early stages of training. The actions during the initial states were τ-dependent as δ_L was equal to 0. The τ-based search resulted in an almost uniform selection for each action, although some τ_{post}-dependent differences existed. The $\max Q$ values increased relatively quickly in RS(λ) and reached that of the initial states. Therefore, the action selection for $\max Q$ and other exploratory action selections were appropriately distributed based on the δ_G.

6.1 Efficient Exploration

Appropriate adjustment of the search volume identified the step numbers per episode (Fig. 4). The normal RS exhibited extremely explorative episodes that required 45 steps. The graph depicted the average number of steps in 500 trials. This is likely to manifest as a regular number of steps, as the search was almost entirely dependent on the confidence level τ.

The RS(λ) remained unaltered in extreme conditions, but decreased monotonically. The action sequences for reaching each goal were established initially by updating the Q(λ) value. The narrowed-down action sequences were switched

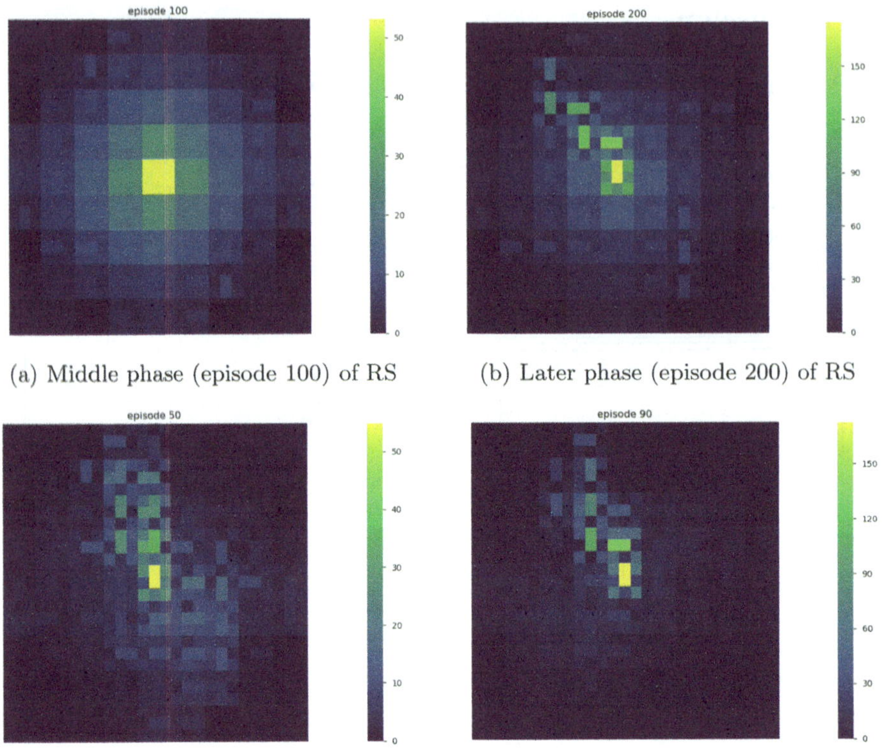

(a) Middle phase (episode 100) of RS

(b) Later phase (episode 200) of RS

(c) Middle phase (episode 50) of RS(λ)

(d) Later phase (episode 90) of RS(λ)

Fig. 5. Heat map of reliability τ for RS and RS(λ). The warmer color indicates more frequent visits. Four corners of the square indicate the mean of the $\tau(s, a)$ values. The center represents the maximal value of all actions.

at key points for accurate value estimation. Owing to this accuracy, the selection ratio of each action series was believed to be appropriately distributed.

Figure 5 depicts the heatmap of the confidence level τ of RS and RS(λ), at the mid and end points of the study. The confidence level for each state in the normal RS increased in the mid and end points. The decay was effective owing to γ_τ, on noticeable distance. However, RS(λ) focused on the rewarding 8 and 7 goals at the mid point of the game, suggesting that value-based exploration was appropriate.

6.2 Return-Altered Policies and Their Behavior

The goal choice of RS(λ) was reward size-dependent, as observed in goal distribution (6(a)).

During the normal RS, E_G began to increase before 100th episode. The top- and bottom-four goals of the reward amount were selected using identical policy.

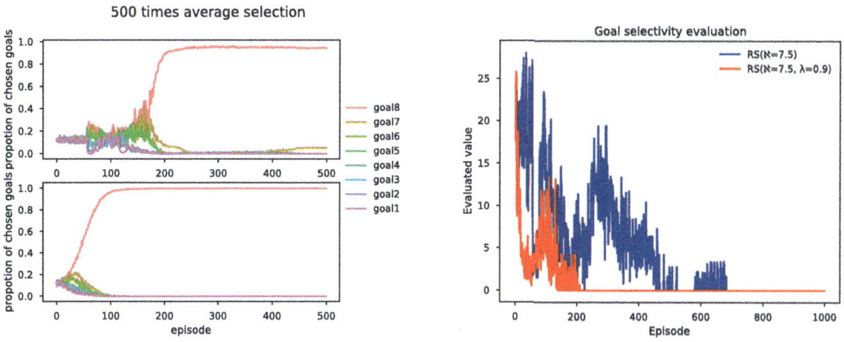

Fig. 6. Analysis of the accomplished goals. **Left**: The goal selection probability. The vertical axis represents the probability, and the horizontal axis represents the elapsed time. **Right**: Time series distribution of the selection distribution scores. The scores were obtained from the difference between the reward value (in descending order) and the reward value (in descending order) of the actual selection percentage. For example, when the actual selection percentage was (8, 7, 5, 6, 3, 4, 1, 2), the score value was 6.

This depicted the percentage of people that chose their destination among the four corners and one of the two goals depending on the reward value.

RS (λ) appears to be a selection proportion, following the reward order. After 50 episodes, δ_G averaged -1.5. When the value of the action series of each goal was assumed to be equal to that of the respective reward, the approximate δ_L of each action series were -1.5, -2.5, -3.5, -4.5, -4.5, -5.5, -6.5, -7.5, and -8.5. Considering the number of selections, where each of these values were equivalent, we obtained 0.21, 0.19, 0.16, 0.14, 0.11, 0.09, 0.06, 0.04 values. This was similar to the selection ratio in Fig. 6(a). Therefore, δ_G approached 0 when E_G increased. The choice enhanced the magnitude of the reward, and the chosen goal converges that of 8. Hence, exploration that appropriately considers the reward is realized.

7 Conclusions

We demonstrate that RS(λ), which couples eligibility trace with RS+GRC, enabled efficient learning toward satisfaction, based on the rewards acquired by the behavior policy. RS(λ) autonomously balanced the improvements through evaluation and exploration of policies. This was based on the returned amount during non-satisfying aspirational levels. Therefore, the search space was reduced, according to the returns and the objective, suggesting that a satisfactory solution could be identified with large search spaces. Here, we did not consider uncertainty and fluctuations of the environment. Thus, the policy evaluation in uncertain and/or fluctuating environments will be investigated further.

References

1. Precup, D., Sutton, R.S., Singh, S.: Eligibility traces for off-policy policy evaluation. In: 17th ICML 2000, pp. 759–766 (2000)
2. Shinriki, M., Wakabayashi, H., Kono, Yu., Takahashi, T.: Flexibility of emulation learning from pioneers in nonstationary environments. In: Ohsawa, Y., et al. (eds.) JSAI 2019. AISC, vol. 1128, pp. 90–101. Springer, Cham (2020). https://doi.org/10.1007/978-3-030-39878-1_9
3. Simon, H.A.: Rational choice and the structure of the environment. Psychol. Rev. **63**(2), 129–138 (1956)
4. Takahashi, T., Kono, Y., Uragami, D.: Cognitive satisficing: bounded rationality in reinforcement learning. Trans. Jap. Soc. AI, **31**(6), AI30-M_1–11 (2016)
5. Tamatsukuri, A., Takahashi, T.: Guaranteed satisficing and finite regret: analysis of a cognitive satisficing value function. Biosystems **180**, 46–53 (2019)
6. Ushida, Y., Kono, Y., Takahashi, T.: Proceedings of JSAI 2017, 4C2-2in2 (2017)

Designing Lightweight Feature Descriptor Networks with Depthwise Separable Convolution

Yeo Ree Wang(✉) and Atsunori Kanemura(✉)

LeapMind Inc., Tokyo 150-0044, Japan
{yeo,atsu-kan}@leapmind.io

Abstract. Extracting feature points and their descriptors from images is one of the fundamental techniques in computer vision with many applications such as geometric fitting and camera calibration, and for this task several deep learning models have been proposed. However, existing feature descriptor networks have been developed with the intention of improving the accuracy, and consideration for practical networks that can run on embedded devices has somewhat been deferred. Therefore, the objective of this study is to devise light feature descriptor networks. To this end, we employ lightweight convolution operations that have been developed for image classification networks (e.g. SqueezeNet and MobileNet) for the purpose of replacing the normal convolution operators in the state-of-the-art feature descriptor network, RF-Net. Experimental results show that the model size of the detector can be reduced by up to 80% compared to that of the original size with only a 11% degradation at worst performance in our final lightweight detector model for image matching tasks. Our study indicates that the modern convolution techniques originally proposed for small image classification models can be effectively extended to designing tiny models for the feature descriptor extraction and matching portions in deep local feature learning networks.

Keywords: Neural networks · Lightweight feature descriptor networks · Depthwise separable convolution · Deep learning

1 Introduction

Detecting salient feature points and their corresponding feature descriptors is a classical and useful technique in computer vision [4], and recent studies [5,8] have proposed high-performing feature descriptors based on deep neural networks. Although the methodology of designing practical networks that can run on embedded devices is one of the active research areas in deep learning [2,3,6,7], the existing feature detection networks incur prohibitively large computational load.

The goal of feature detection is to robustly locate points of interest between similar images. The de facto standard feature descriptor before deep learning

K. Yada et al. (Eds.): JSAI 2020, AISC 1357, pp. 183–191, 2021.
https://doi.org/10.1007/978-3-030-73113-7_17

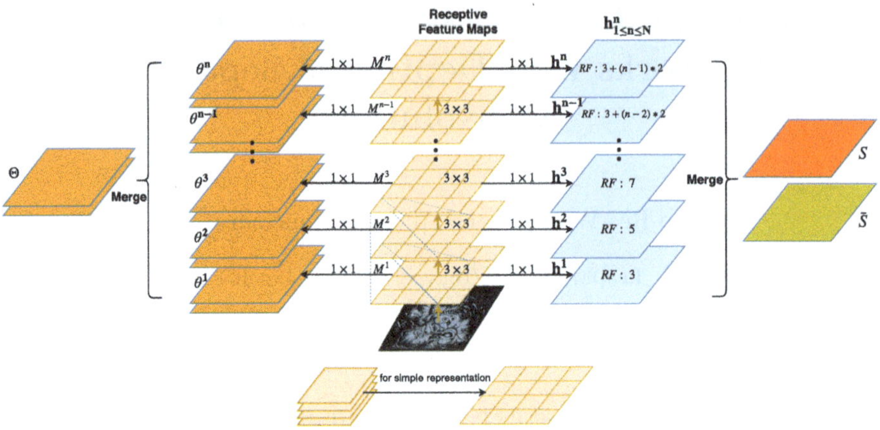

Fig. 1. Visual representation of RF-Net detector (Adopted from [8]).

was SIFT [4] and its variants, which were designed by "hand-crafting" a feature detector and a feature descriptor. Once keypoints are extracted from an image, only the positions of the pixels are known. In order to compare the keypoints, feature descriptors, or in other words, the area or patch of pixels surrounding each detected keypoint, are compared against all of the corresponding patches from another image. The feature descriptors are simply vectors that can be compared using metrics such as the Euclidean distance. A matching pair of keypoints would therefore be defined by the minimum Euclidean distance between two feature descriptor vectors.

Deep learning methods for feature detection and description have recently been proposed. One such example is the state-of-the-art LF-Net, which is "trainable end-to-end and does not require using a hand-crafted detector to generate training data" [5]. LF-Net employs a novel architecture split into two divisions: the keypoint detector/descriptor extractor and the learning of the global image and descriptor patches. However, a few technical drawbacks of LF-Net are that the datasets it uses are quite huge and the model requires camera pose information. A simpler, improved version of LF-Net called RF-Net [8] also achieved state-of-the-art results. The technical improvements of RF-Net are that it offers more receptive fields of an image (Fig. 1), which assists in more effective keypoint detection, and a more stable descriptor training approach [8]. The quality of life changes are that it is a simpler model as it does not have to account for 3D data (i.e. camera orientation and pose) and trains on lighter datasets.

The RF-Net detector is the first step in the network to determine keypoints. Once the detector has found enough keypoints, descriptors, or patches of pixels around each keypoint are then collected and run through the second portion of RF-Net, which is the training network. The detector is where our focus was placed in order to create a lightweight detector model.

In this paper, we investigated the adaptability of RF-Net into a model with less computational load. Specifically, we employ lightweight convolutional layers such as depthwise separable layers to RF-Net's keypoint detection model. We found that decreasing the computational load by 80% of that of the original detector's computational load had only a minimal reduction (at most 11%) on the matching scores of RF-Net. However, the number of detected keypoints decreased depending on the amount of replaced convolutional layers with depthwise separable layers. In addition, when evaluating our final lightweight model on an actual machine with respect to GPU inference runtime percentage breakdown, we found that there were slightly higher percentages of overall pre-processing times but shorter keypoint matching and post-processing percentages in the context of the total runtime when compared to that of [8]'s pretrained model. If the principles employed by state-of-the-art image classification networks such as SqueezeNet [3] and MobileNet [2] are also applicable to feature descriptor networks, then there is potential in making feature descriptor networks available on older machines or edge devices.

2 Making RF-Net Detector Lightweight

2.1 Basic Convolutional Layer Architecture

A standard convolutional layer is fed feature maps of height H, width W, and input channels C_{in}, of which the outputs are then batch normalized, and subsequently passed into one of a variety of possible activation layers. Assuming that the convolutional layer uses a square kernel of size $K \times K$ and that the output channels of the feature maps are C_{out}, the computational cost of a standard convolutional layer is:

$$K^2 C_{\text{in}} C_{\text{out}} HW. \tag{1}$$

2.2 Making RF-Net Detector Lightweight

Our method to reduce the theoretical computational load of the RF-Net detector was done through the replacement of the default convolutional layers with depthwise separable layers. The default RF-Net detector uses ten *concerned receptive fields* to generate feature maps in order to detect points of interests (Fig. 1). The output of each subsequent convolutional layer is convolved with a kernel size of 3×3 with a stride of 1, which increases the receptive field size from 3, 5, 7, ..., 21.

Since the kernel size is not 1×1 for most of the layers shown in the middle column (receptive feature maps) of Fig. 1, this makes the RF-Net detector a good candidate for application of depthwise separable convolution proposed by MobileNet [2]. This portion of RF-Net is where we centered our experiments on. When a convolutional layer is replaced by a depthwise separable layer, it is decomposed into a depthwise layer and a pointwise layer.

In a normal convolutional layer, each area screened by the kernel outputs a value, which then is combined with other input channels to become part of a new feature map. A basic RGB image often has its output channels reduced to 3 from 1. Multiple kernels are then stacked together to output multi-channel feature maps. In short, an image is transformed C_{out} times.

2.3 Estimating Improvements

In a depthwise layer, an image is only transformed once as each kernel iterates only over one channel. As a result, the input channels, C_{in}, and output channels, C_{out}, remain the same. Afterwards, a normal convolutional layer with a kernel size of 1×1 is then applied on the outputs from the depthwise layer. In the same vein as a normal convolutional layer, the number of output channels is determined by the amount of kernels used.

While it may seem that basic convolutional and depthwise separable layers process feature maps in the same way, the depthwise separable layer saves immensely due to the way the transformation of the feature maps is only handled once:

$$K^2 C_{\text{in}} HW. \tag{2}$$

As represented above, there is a factor less of C_{out} that needs to be considered for the depthwise layer. As for the pointwise layer where the kernel size is 1×1, and thus negligible in the computation, the pointwise layer can be represented by:

$$C_{\text{in}} C_{\text{out}} HW. \tag{3}$$

Once summed together, the theoretical computational cost of a depthwise separable layer comes out to:

$$K^2 C_{\text{in}} HW + C_{\text{in}} C_{\text{out}} HW = C_{\text{in}} HW(K^2 + C_{\text{out}}). \tag{4}$$

When the basic convolutional layer is compared to the depthwise separable layer:

$$\frac{K^2 C_{\text{in}} HW}{C_{\text{in}} HW(K^2 + C_{\text{out}})} = \frac{K^2}{(K^2 + C_{\text{out}})}. \tag{5}$$

Thus, the depthwise separable convolutional layer is a factor of $K^2/(K^2+C_{\text{out}})$ smaller compared to that of the basic convolutional layer.

3 Experimental Results

3.1 Configurations

Datasets. We used the same datasets as the RF-Net study [8]: the HPatches [1] and EF [9] datasets. HPatches includes 51 sequences, each of which has one reference image and five images with varying photometric changes. RF-Net [8] chose to split this dataset further into viewpoint (30 pair images) and illumination (285 pair images) categories for testing purposes. The EF dataset consists of five different base images with each base image having corresponding images in various viewpoint and illumination changes.

The Initial Convolutional Layer. After applying depthwise separable convolution to all the layers in the RF-Net detector, we found that it was most effective to not replace the initial convolutional layer (data not shown). By not doing so, the validation and matching scores of the trained models were found to be higher compared to the models in which the initial convolutional layer was replaced. The reason for this may be that the kernel size of the initial convolutional layer was already 1×1, so making this layer depthwise separable may have traded too much feature map information for too little reduction in computational load.

Table 1. Nine variants of lightweght RF-Net architectures.

#Layers replaced (Label)	Replaced Layers
1	self.conv2
2	self.conv2 ~ self.conv3
3	self.conv2 ~ self.conv4
4	self.conv2 ~ self.conv5
5	self.conv2 ~ self.conv6
6	self.conv2 ~ self.conv7
7	self.conv2 ~ self.conv8
8	self.conv2 ~ self.conv9
9	self.conv2 ~ self.conv10

Comparison. We compare the performance of our lightweight RF-Net with those reported in [8] and also those of the pretrained model (downloaded from https://github.com/Xylon-Sean/rfnet). We examined nine variants of lightweight RF-Net architectures shown in Table 1.

3.2 Performance Evaluation of the Lightweight RF-Net Detector

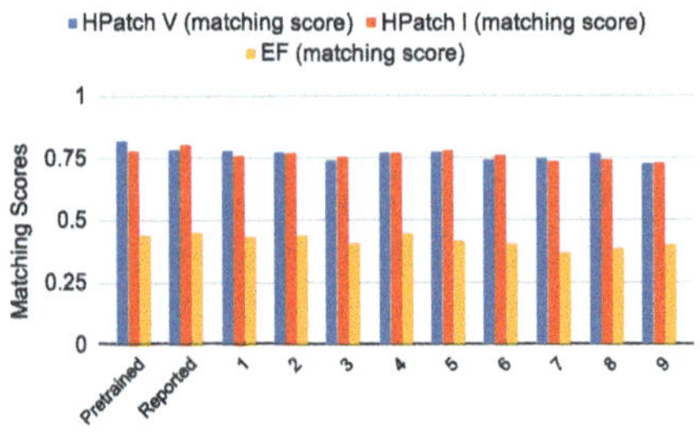

Fig. 2. Model evaluation for our various lightweight RF-Net detector showing the number of layers replaced vs. matching scores.

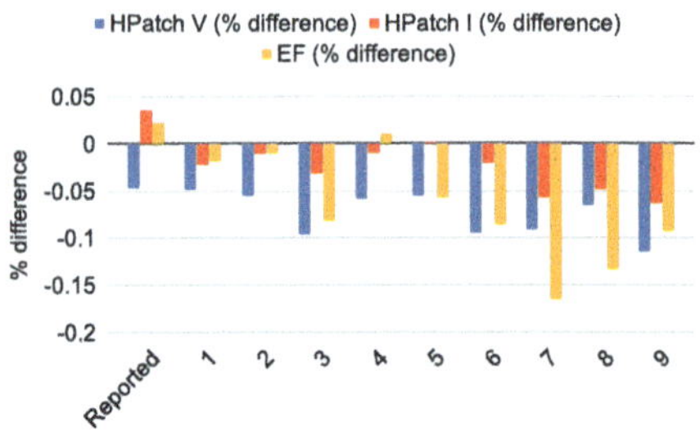

Fig. 3. Percentage difference in matching score when comparing [8]'s pretrained model with our various lightweight RF-Net detector models.

As shown in Fig. 2, when all other layers besides the initial convolutional layer in RF-Net's detector are replaced with lightweight depthwise separable layers, the most single significant drop in matching score (Fig. 3) is by about 18% in EF dataset's matching score when the first six layers after the initial convolutional layer are replaced. However, as shown in Figs. 4 and 5 for our final lightweight model, the worst performance is around 11% for the HPatch viewpoint dataset.

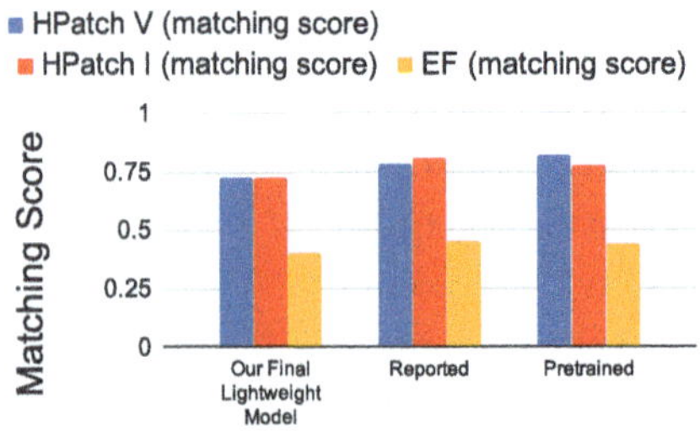

Fig. 4. Model evaluation for our final lightweight RF-Net detector vs. matching scores.

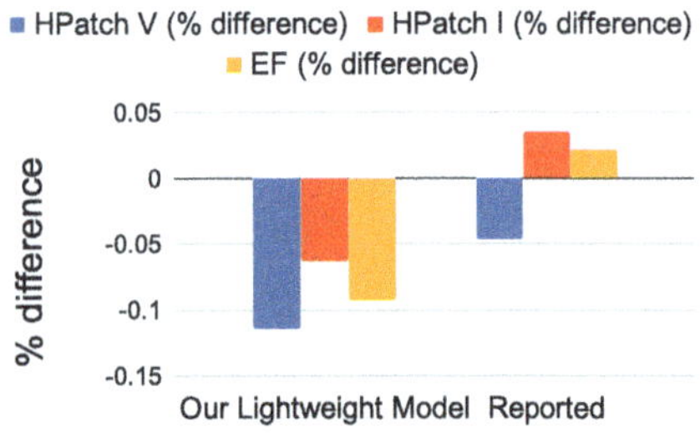

Fig. 5. Percentage difference in matching score when comparing [8]'s pretrained model with our final lightweight RF-Net detector model and [8]'s reported model.

In Fig. 6, when the computational load of the RF-Net detector is slowly reduced by incrementally replacing normal convolutional layer with depthwise separable convolutional layers, the matching scores across all datasets do not vary significantly.

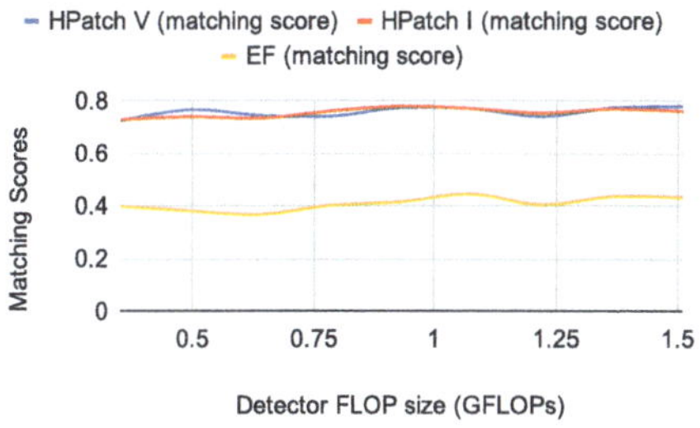

Fig. 6. Lightweight RF-Detector model performance in terms of computational load vs. matching scores. Dimensions of test image are 240 × 320.

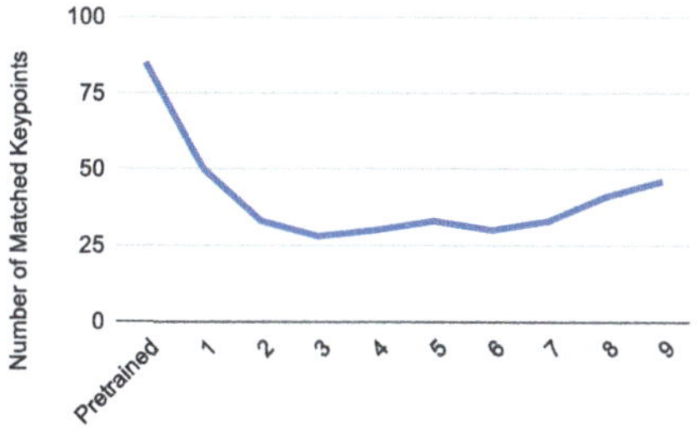

Fig. 7. Matched keypoints vs. number of convolutional layers replaced for author provided sample image (Fig. 8).

Fig. 8. Visual representation (using a sample image provided by [8]) of matched keypoints in [8]'s pretrained model.

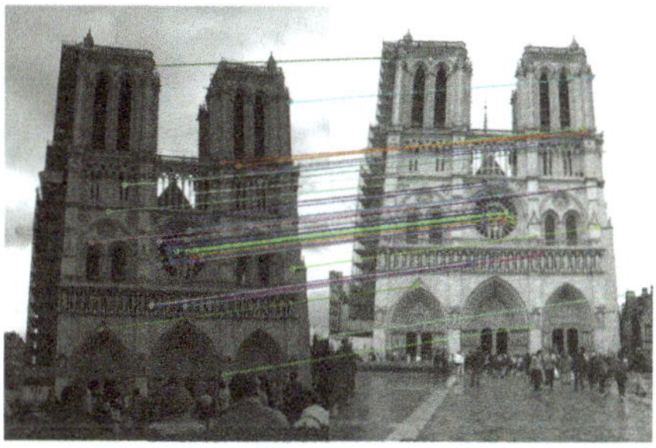

Fig. 9. Visual representation (using a sample image provided by [8]) of matched keypoints in our final lightweight model.

3.3 Computational Load and Number of Matched Keypoints

To measure the amount of keypoints matched, we used the default image that the authors of RF-Net used as an example. As seen in Fig. 7, the number of matched keypoints drops significantly after only one convolutional layer is replaced. After about six convolutional layers are replaced with depthwise separable layers, the number of matched keypoints seems to increase compared to the number of the previous layers. Comparisons between visual representation in Figs. 8 and 9 show roughly how many and which keypoints are lost from making the RF-Net detector lightweight.

3.4 Practical Runtime Efficiency

To empirically determine the effect of our lightweight model on an actual computer, we evaluated the GPU runtimes during inference. Each significant code block within the inference script was placed between two CUDA events, which were then synchronized with the CUDA stream in order to record accurate runtimes. Twenty trials each for [8]'s pretrained model and our own final lightweight model were conducted, and the lowest and highest two total GPU runtimes were removed from the mean. The visual representations in Figs. 10 and 11 show that there were no significant differences between the code blocks within the inference script. However, interesting points that appeared were that our final lightweight model had higher initialization (by 0.1%), detector loading (by 0.1%), descriptor loading (by 0.1%), model loading to GPU times (by 0.9%), and keypoint detection/computation (by 0.1%) in the percentage breakdown when compared with the breakdown of [8]'s pretrained model. However, our final lightweight model had lower breakdown portions for keypoint matching (by 1%) and post-processing (by 0.1%).

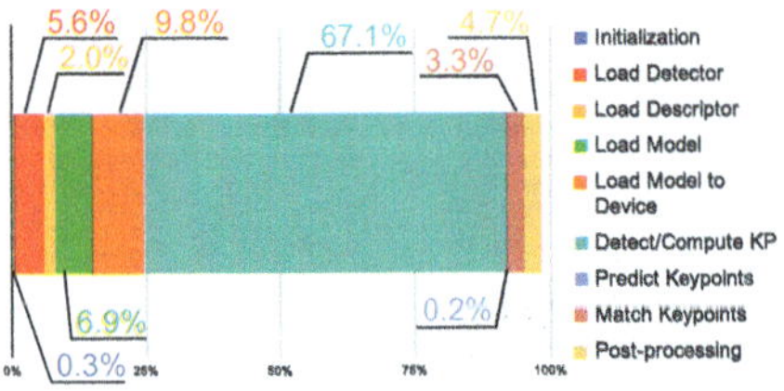

Fig. 10. Percentage breakdown of inference run times (ms) for [8]'s pretrained model. Dimensions of test image are 480 × 640. GPU used was a single NVIDIA GeForce GTX 1080 Ti.

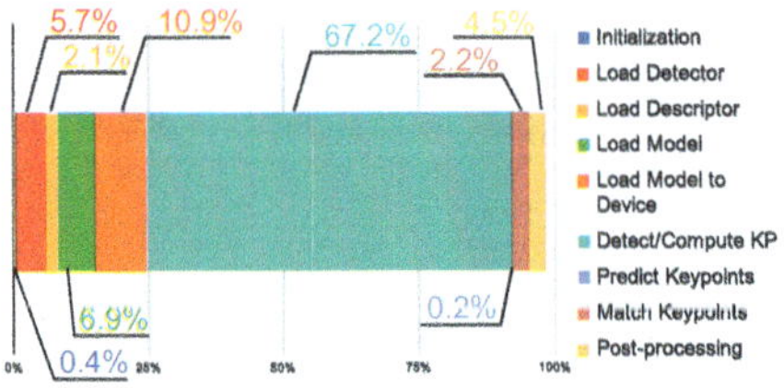

Fig. 11. Percentage breakdown of inference run times (ms) for our final lightweight model. Dimensions of test image are 480 × 640. GPU used was a single NVIDIA GeForce GTX 1080 Ti.

4 Conclusions

In this work, we applied the technique of depthwise separable convolution to RF-Net, a state-of-the-art feature descriptor network. We heavily reduced the computational load of the feature detector portion of the architecture with a negligible change in terms of relative matching scores. In addition, the proposed changes to the detector model of RF-Net revealed that there was only a minor decrease in overall matching score performance when comparing our final lightweight detector model with that of [8]'s pretrained model. While there were small fluctuations in relative matching scores during the process of making convolutional layers in the RF-Net detector lightweight, we found that a more significant trade-off was the decrease in number of matched keypoints.

Further experiments investigating the practical runtime efficiency of both [8]'s pretrained model and our own lightweight model revealed that there was no significant change in the percentage breakdown of the inference runtimes. While there were higher percentages of pre-processing and keypoint detection/computation in the overall scope of the inference runtimes in our final lightweight model compared to that of [8]'s pretrained model, there were also lower percentages in the keypoint matching and post-processing blocks. From the data gleaned in the subsequent experiments, it appears that there is still a gap between theoretical manipulation and pragmatic application. Deep-learning frameworks may not implement depthwise separable layers as efficiently now, and the runtime benchmarking of CUDA events might also not be the best metric for evaluation of inference speed.

References

1. Balntas, V., Lenc, K., Vedaldi, A., Mikolajczyk, K.: HPatches: a benchmark and evaluation of handcrafted and learned local descriptors. In: IEEE/CVF Conference on Computer Vision and Pattern Recognition (CVPR) (2017)
2. Howard, A.G., et al.: MobileNets: efficient convolutional neural networks for mobile vision applications. arXiv:1704.04861 (2017)
3. Iandola, F.N., Han, S., Moskewicz, M.W., Ashraf, K., Dally, W.J., Keutzer, K.: SqueezeNet: AlexNet-level accuracy with 50x fewer parameters and <0.5MB model size. arXiv:1602.07360 (2016)
4. Lowe, D.G.: Distinctive image features from scale-invariant keypoints. Int. J. Comput. Vis. **60**, 91–110 (2004). https://doi.org/10.1023/B:VISI.0000029664.99615.94
5. Ono, Y., Trulls, E., Fua, P., Yi, K.M.: LF-Net: learning local features from images. In: Advances in Neural Information Processing Systems (NIPS) (2018)
6. Sandler, M., Chu, G., Chen, L.-C.: Searching for MobileNetV3. Presented at the (2019)
7. Sandler, M., Howard, A., Zhu, M., Zhmoginov, A., Chen, L.C.: MobileNetV2: inverted residuals and linear bottlenecks. In: IEEE/CVF Conference on Computer Vision and Pattern Recognition (CVPR) (2018)
8. Shen, X., et al.: RF-Net: an end-to-end image matching network based on receptive field. In: IEEE/CVF Conference on Computer Vision and Pattern Recognition (CVPR) (2019)
9. Zitnick, C.L., Ramnath, K.: Edge foci interest points. Presented at the (2011)

Hybrid Quantum-Classical Dynamic Programming Algorithm

Chih-Chieh Chen[1(✉)], Kodai Shiba[1,2], Masaru Sogabe[1], Katsuyoshi Sakamoto[2], and Tomah Sogabe[2(✉)]

[1] Grid Inc., Tokyo, Japan
chen.chih.chieh@gridsolar.jp
[2] The University of Electro-Communications, Tokyo, Japan
sogabe@uec.ac.jp

Abstract. The Markov decision process has various applications in engineering, economics, operations research and artificial intelligence. Quantum computers provide a new way to tackle the computational problems in solving Markov decision process. We develop quantum circuits for dynamic programming algorithm to solve for the Markov decision process. The heuristic circuit construction method based on linear combinations of unitaries is demonstrated. The matrix decomposition and sampling are discussed. The computability advantage over classical Birkhoff-von Neumann method is proved. The connection to traditional dynamic programming algorithm is discussed in terms of functional equations. Our work suggests a new hybrid quantum-classical approach to dynamic programming algorithms.

Keywords: Dynamic programming · Hybrid quantum-classical algorithm

1 Introduction

The Markov decision process (MDP) is a stochastic model for decision making and optimal control. MDP has applications in engineering, economics, operations re-search, and artificial intelligence. MDP can be solved by dynamic programming and reinforcement learning algorithms [2,6,8,13]. In this work, we use quantum circuit heuristics in dynamic programming algorithms for MDP problems [3]. We provide some formulae for quantum circuit constructions and discuss the matrix decompositions. We then prove that our quantum heuristic method is Birkhoff-von Neumann uncomputable. We also demonstrate an application to a probabilistic grid world example, and we discuss the connection to classical dynamic programming algorithms.

2 Method

A Markov decision process (MDP) is defined by a tuple (S, A, P, R, γ) where S is the set of possible states, A is the set of possible actions, $P^{a}_{ss'}$ is the transition

K. Yada et al. (Eds.): JSAI 2020, AISC 1357, pp. 192–199, 2021.
https://doi.org/10.1007/978-3-030-73113-7_18

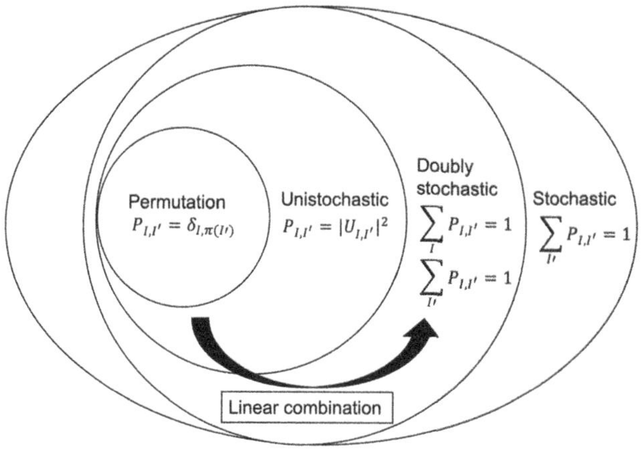

Fig. 1. Set diagram for subsets of stochastic matrices

matrix, $R^a_{ss'}$ is the reward function, and γ is the discount rate. The goal is to find the policy which maximize the cumulative expected future rewards [2,6,8,13]. The problem can be solved by solving Bellman's functional equation for the value function

$$V_s \leftarrow \max_a \{\sum_{s'} P^a_{ss'}[R^a_{ss'} + \gamma V_{s'}]\}$$

or for the state-action Q-function

$$Q(s,a) \leftarrow \frac{1}{L}\sum_{j=1}^{L}\{R^a_{ss'_j} + \gamma \max_{a'} Q(s'_j, a')\}$$

where samples are drawn from $s'_j \sim P^a_{ss'}$. In the general MDP, the transition matrices are stochastic matrices. There are some subsets of stochastic matrices that we can consider. Permutation matrices form a subset of unistochastic matrices. Unistochastic matrices form a subset of doubly stochastic matrices. Doubly stochastic matrices form a subset of stochastic matrices. The relation among these subsets is depicted in Fig. 1.

The following theorem due to Birkhoff and von Neumann is taken from Reference [10].

Theorem 1 (Birkhoff-von Neumann). *An N by N matrix P is doubly stochastic if and only if there are N by N permutation matrices $\pi_1, \pi_2, ..., \pi_k$ and positive real numbers $t_1, t_2, ..., t_k$ such that $t_1 + t_2 + ... + t_k = 1$ and $P = t_1\pi_1 + t_2\pi_2 + ... + t_k\pi_k$. The order of the decomposition k satisfies $k \leq N^2 - N + 1$.*

The constructive proof given in Reference [10] gives the form

$$P = \alpha_1\pi_1 + (1-\alpha_1)\alpha_2\pi_2 + ... + (1-\alpha_1)(1-\alpha_2)...(1-\alpha_k)\pi_k.$$

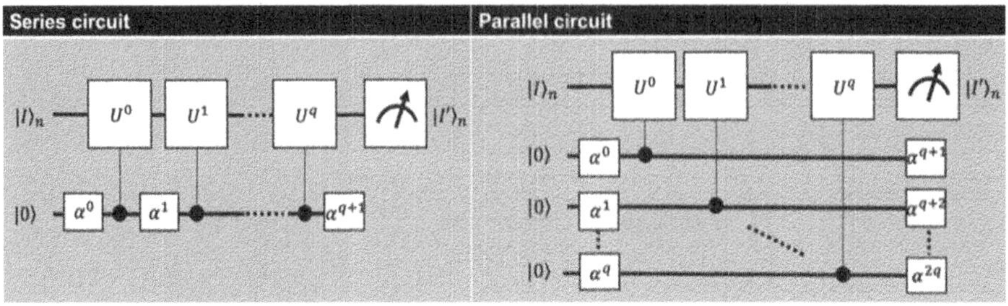

Fig. 2. Parallel and series circuits combinations for unitaries

It is straightforward to sample this Birkhoff-von Neumann decomposition by classical computers or quantum circuits, provided that we have the implementations of the permutations [12]. The worst-case complexity of the decomposition costs $O(N^2)$, but there are greedy heuristics to obtain better decompositions [7]. However, the computational complexity to sample the decomposition is the same for both classical and quantum circuits. Hence the worst-case scenario is that we can use quantum circuits to do what classical computers can do, with the same efficiency. The question is that whether we can find a subset of stochastic matrices such that quantum circuits have some advantages. The sampling of a single unistochastic matrices is known [11]. To solve for more general problems, one question is how to sample doubly stochastic matrices that are not unistochastic? Linear combination of unitaries provides a possible way to construct matrices [5]. Given some unitaries or permutations $U^0, U^1, \ldots, U^q$, we consider two simple ways to combine them: series and parallel circuits. Figure 2 shows the circuits and corresponding transition matrices.

The transition matrix given by series circuit is

$$P_{I,I'} = \sum_{k} | \sum_{c_0,c_1,\ldots,c_q} \alpha^{q+1}_{k,c_q} \ldots \alpha^{1}_{c_1,c_0} \alpha^{0}_{c_0,0} \langle I' | (U^q)^{c_q} \ldots (U^1)^{c_1} (U^0)^{c_0} | I \rangle |^2.$$

The transition matrix given by parallel circuit is

$$P_{I,I'} = \sum_{k_0,k_1,\ldots,k_q} | \sum_{c_0,c_1,\ldots,c_q} \alpha^{2q}_{k_q,c_q} \alpha^{q}_{c_q,0} \ldots \alpha^{q+2}_{k_1,c_1} \alpha^{1}_{c_1,0} \alpha^{q+1}_{k_0,c_0} \alpha^{0}_{c_0,0} \langle I' | (U^q)^{c_q} \ldots (U^1)^{c_1} (U^0)^{c_0} | I \rangle |^2.$$

These constructions are easy to sample using quantum circuits, but difficult to sample using classical circuits. To proof that these simple combinations can sample matrices beyond unistochastic matrices, and to illustrate how the decomposition works in practice, we will construct an explicit circuit example. We will use the following lemma. This lemma is a simple extension of an example given in Reference [10].

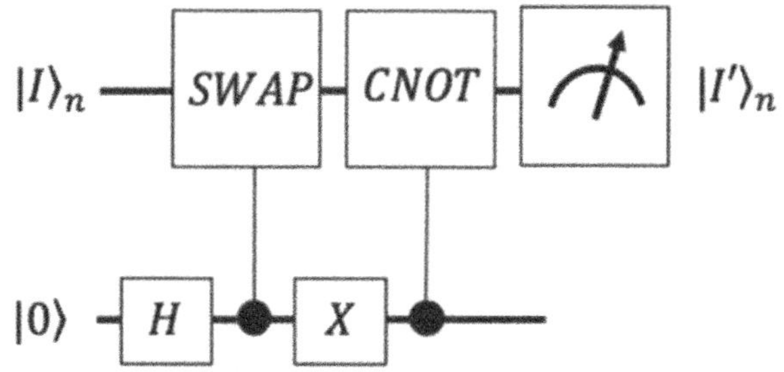

Fig. 3. Series quantum circuit for a non-unistochastic matrix

Lemma 1. *The matrix*

$$P = \begin{pmatrix} 1 & 0 & 0 & 0 \\ 0 & \frac{1}{2} & \frac{1}{2} & 0 \\ 0 & \frac{1}{2} & 0 & \frac{1}{2} \\ 0 & 0 & \frac{1}{2} & \frac{1}{2} \end{pmatrix}$$

is doubly stochastic but not unistochastic.

Proof. Assuming P is unistochastic. Then there exists an unitary matrix u such that $P_{i,j} = |u_{i,j}|^2$. Without loss of generality, the unitary matrix can be expressed as

$$u = \begin{pmatrix} e^{i\theta_{00}} & 0 & 0 & 0 \\ 0 & \frac{1}{\sqrt{2}}e^{i\theta_{11}} & \frac{1}{\sqrt{2}}e^{i\theta_{12}} & 0 \\ 0 & \frac{1}{\sqrt{2}}e^{i\theta_{21}} & 0 & \frac{1}{\sqrt{2}}e^{i\theta_{23}} \\ 0 & 0 & \frac{1}{\sqrt{2}}e^{i\theta_{32}} & \frac{1}{\sqrt{2}}e^{i\theta_{33}} \end{pmatrix},$$

where θ_{ij} are real numbers. Computing the product gives

$$u^\dagger u = \begin{pmatrix} 1 & 0 & 0 & 0 \\ 0 & 1 & \frac{1}{\sqrt{2}}e^{i(\theta_{12}-\theta_{11})} & . \\ 0 & . & . & . \\ 0 & . & . & . \end{pmatrix}.$$

One of the off-diagonal elements is impossible to be zero, hence u is not unitary. The claim is proofed by contradiction. □

Lemma 2. *The series circuit can be used to sample some doubly stochastic matrix which is not unistochastic.*

Proof. Consider a Birkhoff-von Neumann decomposition of the matrix:

$$\begin{pmatrix} 1 & 0 & 0 & 0 \\ 0 & \frac{1}{2} & \frac{1}{2} & 0 \\ 0 & \frac{1}{2} & 0 & \frac{1}{2} \\ 0 & 0 & \frac{1}{2} & \frac{1}{2} \end{pmatrix} = \frac{1}{2}\begin{pmatrix} 1 & 0 & 0 & 0 \\ 0 & 0 & 1 & 0 \\ 0 & 1 & 0 & 0 \\ 0 & 0 & 0 & 1 \end{pmatrix} + \frac{1}{2}\begin{pmatrix} 1 & 0 & 0 & 0 \\ 0 & 1 & 0 & 0 \\ 0 & 0 & 0 & 1 \\ 0 & 0 & 1 & 0 \end{pmatrix}.$$

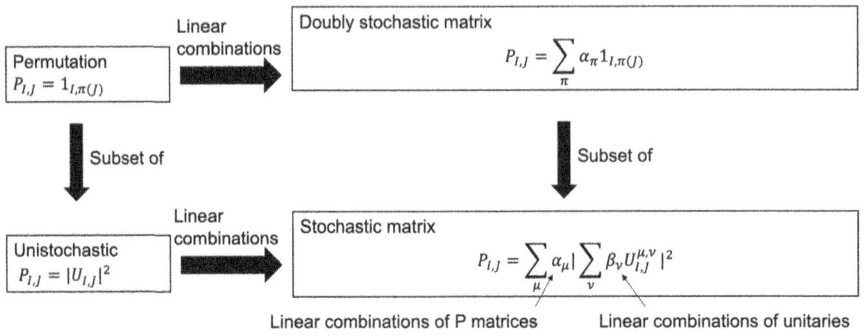

Fig. 4. Relation between classical and quantum combinations

By Lemma 1, the left-hand-side is not unistochastic. The permutations at the right-hand-side are simply the SWAP gate and the CNOT gate. Hence the matrix can be sampled by using the quantum circuit in Fig. 3. □

In general, we can construct linear combinations of the form:

$$P_{I,J} = \sum_{\mu} \alpha_\mu | \sum_{\nu} \beta_\nu U_{I,J}^{\mu,\nu} |^2.$$

The linear combination of unitaries inside the square has no obvious classical efficient method to sample. Expanding the absolute-square, the terms can be collected into a linear combination of unistochastic matrices and a linear combination of Hadamard products between unitary matrices. The relation between quantum combination and classical combination is illustrated in Fig. 4. The Hadamard product terms in quantum combinations do not exist in the classical Birkhoff-von Neumann decomposition. These extra terms provide extra resources for computations. The following theorem shows that the quantum combination is Birkhoff-von Neumann uncomputable.

Theorem 2. *The series circuit can be used to sample some stochastic matrix which is not doubly stochastic.*

Proof. Consider the circuit in Fig. 5, which is modified from the circuit in Fig. 3 by changing the X gate to H gate. The corresponding P matrix is

$$P = \begin{pmatrix} 1 & 0 & 0 & 0 \\ 0 & \frac{1}{2} & \frac{1}{4} & \frac{1}{4} \\ 0 & \frac{1}{2} & \frac{1}{4} & \frac{1}{4} \\ 0 & 0 & 0 & 1 \end{pmatrix}.$$

This matrix is not doubly stochastic, and hence cannot be sampled by Birkhoff-von Neumann decomposition. □

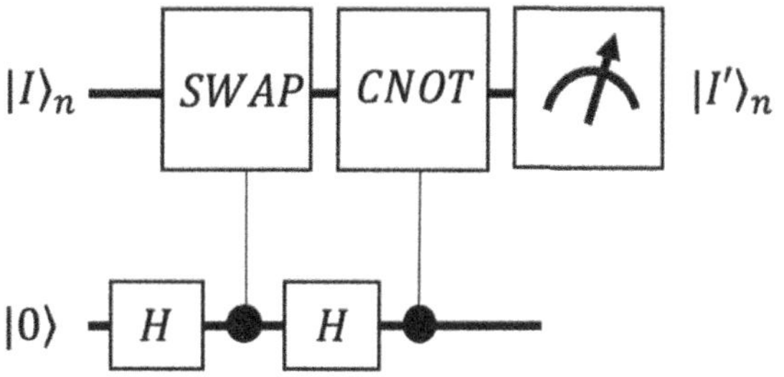

Fig. 5. Series circuit for non-doubly stochastic matrix

We implement a simple grid world example to illustrate the proposed algorithm in its classical limit [1]. The agent moves on a four by four grid with periodic boundary condition. There are five possible actions $A = \{0, \rightarrow, \downarrow, \leftarrow, \uparrow\}$, where 0 means no movement. For each action there is 50% probability that the agent moves to the nearest neighbor. There is 25% probability for the agent being disturbed by one grid to the left and 25% probability for the agent being disturbed by one grid to the right. The permutations π in this case are just add one (mod 4) adder circuit and its inverse. The choice of actions can be controlled by rotation gates on the coin registers. The transition matrix is

$$P^a_{s,s'} = \sum_{k_0,k_1,k_2,k_3} |u_{k_0,0}(\theta_0)|^2 |u_{k_1,0}(\theta_1)|^2 |u_{k_2,0}(\theta_2)|^2 |u_{k_3,0}(\theta_3)|^2 |\langle s'|\pi_3^{k_3}\pi_2^{k_2}\pi_1^{k_1}\pi_0^{k_0}|s\rangle|^2.$$

The transition model is depicted in Fig. 6 and the circuit is shown in Fig. 7.

Bellman formulates general dynamical programming problems in terms of functional equations [2]. One special case is Dijkstra's shortest path algorithm [6]. The functional equation for Dijkstra's algorithm is

$$f_j = \min_i \{f_i + D(i,j)\}.$$

The $D(i,j)$ denotes the distance between node i and node j. This is a deterministic functional equation. It can be considered as a special case of probabilistic Markov decision process

$$f_i = \max_a \{R^a_i + \gamma \sum_j P^a_{i,j} f_j\}.$$

by setting the reward function to be negative distance and setting the transition to be deterministic. The hybrid classical-quantum dynamic programming method is to quantize the transition matrix on a graph to get some advantages. On the other hand, in the model-free reinforcement learning algorithms the transition matrix itself is actually not given. Hence for reinforcement learning applications, it is more promising to look for the quantization of the Q-function [4,9].

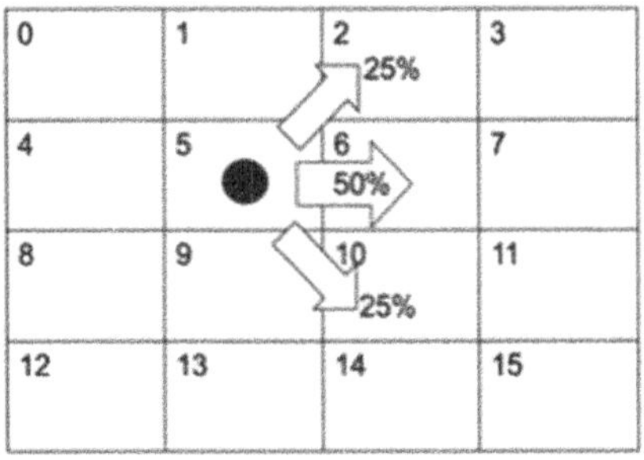

Fig. 6. A probabilistic grid world example

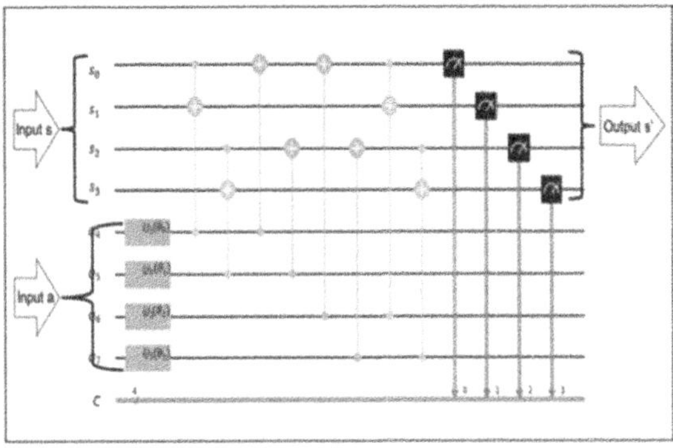

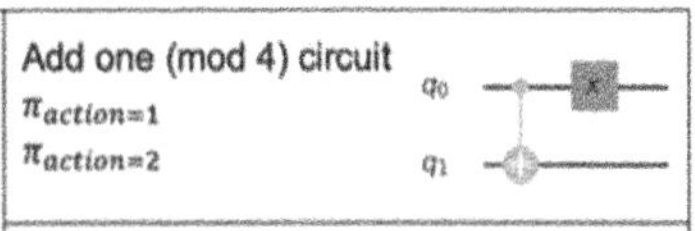

Input angles for different actions					
	a=0	a=1	a=2	a=3	a=4
θ_0	0	π	$\pi/2$	0	$\pi/2$
θ_1	0	$\pi/2$	π	$\pi/2$	0
θ_2	0	0	$\pi/2$	π	$\pi/2$
θ_3	0	$\pi/2$	0	$\pi/2$	π

Fig. 7. Quantum circuit construction

3 Conclusion

In this work, we develop and demonstrate the quantum heuristic circuit constructions for dynamic programming algorithm. There are questions to be answered and future works to be done. We show simple constructions using series and parallel gates combinations for coin qubits. But there are also other possible circuits to be explored for the coin qubits space. While our constructions cost only polynomial time using quantum circuits and they seem to be difficult for classical circuits to sample, we did not show that the quantum circuit violates any complexity lower bound of classical probabilistic machines. It is not clear neither if there is a general way to quantize Birkhoff-von Neumann decomposition to get quantum advantages. These questions are to be answered in the future works.

Acknowledgements. We thank Naoki Yamamoto for valuable discussions.

References

1. Abraham, H., et al.: Qiskit: an open-source framework for quantum computing (2019). https://doi.org/10.5281/zenodo.2562110
2. Bellman, R.E.: Dynamic Programming. Dover Publications Inc., New York (2003)
3. Chen, C.C., Shiau, S.Y., Wu, M.F., Wu, Y.R.: Hybrid classical-quantum linear solver using noisy intermediate-scale quantum machines. Sci. Rep. **9**, 16251 (2019). https://doi.org/10.1038/s41598-019-52275-6
4. Chen, S.Y.C., Huck Yang, C.H., Qi, J., Chen, P.Y., Ma, X., Goan, H.S.: Variational quantum circuits for deep reinforcement learning. arXiv e-prints arXiv:1907.00397 (2019)
5. Childs, A.M., Wiebe, N.: Hamiltonian simulation using linear combinations of unitary operations. Quantum Info. Comput. **12**(11–12), 901–924 (2012)
6. Denardo, E.V.: Dynamic Programming: Models and Applications. Prentice Hall PTR, Hoboken (1981)
7. Dufossé, F., Uçar, B.: Notes on Birkhoff-von Neumann decomposition of doubly stochastic matrices. Linear Algebra Appl. **497**, 108–115 (2016). https://doi.org/10.1016/j.laa.2016.02.023. http://www.sciencedirect.com/science/article/pii/S0024379516001257
8. Geramifard, A., Walsh, T.J., Tellex, S., Chowdhary, G., Roy, N., How, J.P.: A tutorial on linear function approximators for dynamic programming and reinforcement learning. Found. Trends Mach. Learn. **6**(4), 375–451 (2013). https://doi.org/10.1561/2200000042
9. Hamagami, T., Shibuya, T., Shimada, S.: Complex-valued reinforcement learning. In: 2006 IEEE International Conference on Systems, Man and Cybernetics, vol. 5, pp. 4175–4179 (2006). https://doi.org/10.1109/ICSMC.2006.384789
10. Horn, R.A., Johnson, C.R.: Matrix Analysis, 2nd edn. Cambridge University Press, Cambridge (2012)
11. Shende, V.V., Bullock, S.S., Markov, I.L.: Synthesis of quantum-logic circuits. IEEE Trans. Comput. Aided Des. Integr. Circuits Syst. **25**(6), 1000–1010 (2006). https://doi.org/10.1109/TCAD.2005.855930
12. Shende, V.V., Prasad, A.K., Markov, I.L., Hayes, J.P.: Synthesis of reversible logic circuits. IEEE Trans. Comput. Aided Des. Integr. Circuits Syst. **22**(6), 710–722 (2003). https://doi.org/10.1109/TCAD.2003.811448
13. Sutton, R.S., Barto, A.G.: Reinforcement Learning: An Introduction. A Bradford Book, Cambridge (2018)

A Shift from Material to Informational Aspects of Shared Resources can Promote Language Evolution

Hiroto Yonenoh(✉), Reiji Suzuki, and Takaya Arita

Nagoya University, Furocho, Chikusa Ward, Nagoya, Aichi 464-8601, Japan
yonenoh.hiroto@a.mbox.nagoya-u.ac.jp

Abstract. We aim to investigate how a shift from material to informational aspects of shared resources can promote language evolution using a computer simulation. We focus on the two levels of coevolution: that of biological and cultural evolution, and that of two aspects of linguistic abilities (cognitive and communicative). We also extend our integrated model of gene-culture coevolution of language mainly by incorporating the resource finiteness (e.g., high: limited food resources; low: replicable beneficial information). The experiments show that a decrease in the resource finiteness could promote the language evolution while decreasing their linguistic diversity in terms of their characteristic variation.

Keywords: Genetic and cultural evolution of language · Cognitive and communicative linguistic ability · Resource finiteness

1 Introduction

Humankind is the only animal who can use sophisticated languages with complex grammar and high expressiveness. However, human languages that hominid ancestors used in ancient times are expected to be unsophisticated like animal communications we observe at the present time, and evolved to full languages. How and why did only human languages become so sophisticated? We have been focusing on the following unique characteristics of human languages to find a clue to answer this question: an interplay of the biological evolution of the language ability and the cultural evolution of language [1], and co-creative aspects of linguistic communications emerging from synergy between cognitive (e.g., building language hierarchy) and communicative (e.g., intention sharing) abilities of language [2]. We recently showed that individual and social learning can contribute to the evolution of a co-creative aspect of linguistic communication [3].

In this paper, we further examine the impacts of a shift from ancestral to modern human societies on language evolution, from a social shared resource type perspective. According to [4], human (proto) language system is thought to have initially evolved in an ecological niche called "confrontational scavenging" niche, where it is important to form a group and to cooperate with the group

K. Yada et al. (Eds.): JSAI 2020, AISC 1357, pp. 200–207, 2021.
https://doi.org/10.1007/978-3-030-73113-7_19

members via linguistic communications in order to acquire more food resources (animal carcasses that were killed by another predator or that died of natural causes) by fighting with the other groups. In confrontational scavenging niche, the target resources shared within a group were themselves finite, in the sense that they had to be divided among individuals competitively. However, as human (proto) language system developed, it might have enabled us to share more complex information (e.g., knowledge, invention, and rules) to be shared, and such complex information itself could be adaptive resources. Such informational resources are less finite than material ones or could be even infinite. Our modern societies are full of such infinitely replicable informational resources. We believe that such a shift from material to informational aspects of shared resources is a defining factor that affected the coevolutionary process of cognitive and communicative abilities of language.

We thus aim to understand how the finiteness of the shared resources can affect linguistic evolution. For this aim, we firstly extend integrated models of gene-culture coevolution of language mediated by phenotypic plasticity [5], by incorporating a parameter called "resource finiteness" into resource sharing processes with cognitive and communicative abilities of individuals in a two-dimensional linguistic space where languages and individuals evolve. This parameter determines the degree of finiteness when a resource is distributed to individuals. In the resource sharing process, we also make some other assumptions. We then conduct evolutionary experiments with different settings of the resource finiteness and compare the results.

2 Model

2.1 The Linguistic Space

There are N agents in a population, and they can communicate with each other using their shared languages. Agents and languages exist in a two-dimensional linguistic space represented as a polar coordinate system, as shown in Fig. 1. Each language (L) is defined as a point in the space. The radial distance $r_L \in [0.0, \infty)$

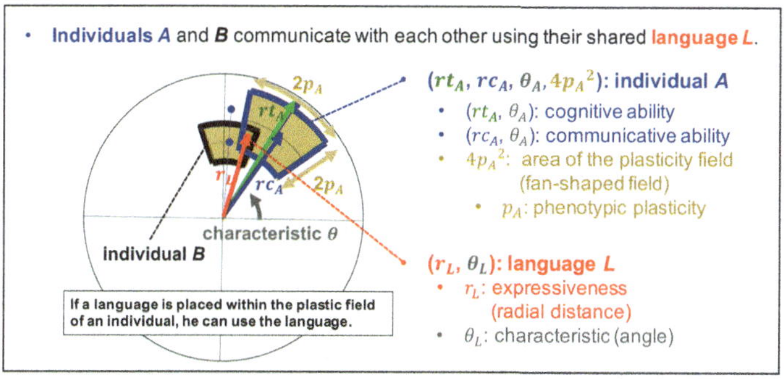

Fig. 1. Representation of individuals and languages in our model.

represents the expressiveness of the language, which contributes to a successful communication using that language. Each agent (A) is represented as two points sharing the same angle. These points, which are respectively represented as (rc_A, θ_A) and (rt_A, θ_A) represent the communicative and cognitive aspects of innate linguistic abilities of the agent, respectively. They are determined by the genotypes $rc_A, rt_A \in [0.0, \infty)$ and $\theta_A \in [0.0, 2\pi)$. rc_A and rt_A, represent the strengths of the agent's communicative and cognitive linguistic abilities, respectively. θ_A, represents the characteristic of the agent's linguistic abilities that enable agents to use languages that have similar characteristics. Each agent also has a fan-shaped field surrounding the point of his communicative linguistic ability, (rc_A, θ_A). The field represents his linguistic plasticity (i.e., the range of his available languages) defined as $2p_A \times 2p_A$ fan-shaped field determined by his genotype $p_A \in [0.0, rc_A]$. We adopted this polar coordinate system to represent a situation that as the expressiveness of a language increases, the space of possible linguistic structures grows, and thus agents with more expressive languages will be harder to communicate successfully with, due to the limited size of the plasticity field.

2.2 Linguistic Interactions

The fitness of an individual A is defined by the equation in Fig. 2, which is composed of the six components with the following coefficient: $\alpha, \beta, \gamma, \delta, \zeta \in [0.0, \infty)$, $\epsilon \in [0.0, 1.0]$. We assume that an agent A can obtain a fitness contribution from the use of each language i that can be used by A (i.e., the all languages within his plasticity field). A1-A2, B1-B2, C, D reflect the higher expressiveness of their shared language (A1) and the larger variation in the characteristics of linguistic abilities (θ_{ij}) among agents communicating in language i (A2) contribute to the larger benefit.

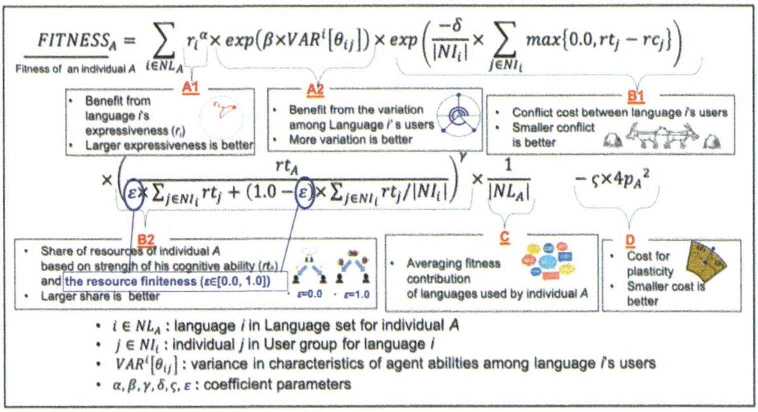

Fig. 2. Fitness function of an agent A.

This group-level benefit is shared by the participants of the group, depending on the resource finiteness and the competitive resource distribution based on their cognitive abilities, which determine the fitness contribution from the language i to A. We assume that the fitness contribution to each agent is distributed in proportional to his cognitive ability (the numerator of B2). This reflects Machiavellian intelligence hypothesis [6], which posits that cognitive abilities of humans have evolved through social competition, where an individual who has more intelligence in order to survive in the game of politics in social groups is more adaptive.

Also, the parameter ϵ determines the finiteness of the shared resource (B2), which is the most important assumption for our research purpose. The sum of their contribution will be equal to the total benefit from the resource when the finiteness is the largest ($\epsilon = 1$). Here, $\epsilon = 1$ represents that the resource is finite (e.g., finite food resources). In this case, individual benefits can decrease as the number of sharing individuals increases. In contrast, when $\epsilon = 0$, the resource is infinite (e.g., beneficial and replicable knowledge), and thus the fitness contribution arising from sharing the resource is not affected by the number of sharing individuals. ϵ is in between these two extreme cases. Furtheremore, we assume that competitive interactions among agents yield the decrease in the total benefit (B1), of which amount is proportional to the averaged strength difference between the communicative and cognitive abilities for every agent j using language i. The final fitness of each agent is the value of the average fitness contribution of his all available languages (C) minus the biological cost for maintaining larger plasticity (D).

Overall, this fitness definition reflects an adaptive environment where the linguistic interactions with the other individuals using the cognitive and communicative faculties are essential to make cooperative resource acquisitions and competitive resource distributions.

2.3 Evolution

The individual population evolves as follows: 1) Parent agents for the next generation are selected using a roulette wheel selection (i.e., the probability with which an agent is picked as a parent is proportional to his fitness). A mutation occurs with a probability P_m for each individual's each genotype, which adds a small random value $R(0.01)$ to the original value of rc_A, rt_A, and p_A; and $R(0.01/rc_A)$ to θ_A, where a function $R(x)$ returns a random number between $-x$ and x with a triangular probability distribution with the mode 0. If the mutated value of p_A exceeds its domain, that process is canceled and another mutation process is applied to the original value. Note that the range of a random value for θ_A is inversely proportional to rc_A of the parent.

On the other hand, the language population evolves the following four cultural processes. Extinction: Any language that was not used by any agents in this generation is removed from the linguistic space. We chose this setting for the sake of simplicity while we believe that it would not generate at least a qualitative difference in the results to assume that languages temporarily without users could survive like the languages in the real world. Cultural change: Each agent creates, for every language she used, an attraction force (vector) F that drags the language toward the agent's location. The length of F is f_{max}/NL_A, where f_{max} is the parameter that determines the maximum length of F, and NL_A is the number of languages within the plasticity field of the agent A. Each language moves to the location determined by the resultant vector of all the forces exerted on it. Division: A language is divided into two languages if the forces of cultural change strongly pull it in opposing directions. We divided received forces into two groups using the second component of PCA [7] of their vectors as a boundary. A division process occurs when the length of either or both resultant forces of the groups exceed the threshold parameter Div_f. These resultant forces determine the locations of two languages resulting from the division. Fusion: When the distance between two languages is close enough, these languages are united into one language.

3 Results and Discussion

We conducted evolutionary experiments for 5000 generations and visualized the results in the linguistic space. The following parameters were used: $N = 1000$, $\alpha = 1.0$, $\beta = 0.5$, $\gamma = 0.05$, $\delta = 2.5$, $\epsilon = 0.0, 1.0$, $\zeta = 1.0$, $P_m = 0.05$, $Div_f = f_{max} = 0.00003$, and $T1 = T2 = 0.001$. The initial value of M was set to 30. Also, the initial values of rc_A, rt_A, and r_L were picked at random from $[0.0, 0.001]$, while the initial values of θ_A and θ_L were set to 0.0.

3.1 Comparison Among the Different Resource Finiteness

First, for each generation in each trial when $\epsilon = 0.0$ and $\epsilon = 1.0$, we examined the averages of (i) the strengths of communicative linguistic abilities of agents ($ComA$), (ii) the expressiveness of languages ($ExpL$), (iii) the strengths of the cognitive linguistic abilities of agents ($CogA$); the variances of (iv) the characteristics of the linguistic abilities of agents ($VarCA$), (v) the characteristics of languages ($VarCL$). Then, we averaged them over the last 500 generations in each trial. The points in Fig. 3 show the distribution of these indices over 20 trials, and the blue and red lines represent their averaged and medium values, respectively. The marks † and * mean that the differences show the statistical significance based on both the paired t-test and Wilcoxon signed-rank test with $\alpha = 0.10$ and 0.05, respectively. We can see that $ComA$, $CogA$, and $ExpL$ tended to evolve to be larger values when the resource finiteness was small ($\epsilon = 0.0$) (i, ii, iii). This means that the smaller resource finiteness can facilitate the overall linguistic evolution processes. At the same time, $VarCA$ and $VarCL$ evolved to be small when $\epsilon = 0.0$ (iv, v). These results imply that a shift from material ($\epsilon = 1.0$) to informational ($\epsilon = 0.0$) aspect of shared resources might have facilitated the adaptive evolution of cognitive and communicative linguistic ability and the expressiveness of language, while decreasing the linguistic diversity.

3.2 Cyclic Coevolution Process of Language and Individual Linguistic Abilities

Subsequently, we examined the time evolution of our model. Figure 4(a) and (b) represent the time evolution of the scaled indices in Fig. 3 as well as (vi) the average of linguistic group size (Average size in the sets of agents who share a same language: $Gsize$), (vii) the number of languages ($NumL$), (viii) the average of phenotypic plasticity of agents (PP), over the first 2000 generations in typical trials when $\epsilon = 0.0$ and $\epsilon = 1.0$, respectively. Note that some time series in these figures are scaled. In both cases, we found a cyclic coevolution process of languages and linguistic abilities. This process is shown in Fig. 5.

First, between around the 975th and around the 1250th generations (1) in Fig. 4(a), the averaged phenotypic plasticity and linguistic group size increased gradually. This seems to be because most individuals had tended to evolve to communicate with as many other individuals as possible by having increased their phenotypic plasticities. On the other hand, during this process (1), the increasing speed of the cognitive linguistic abilities started to increase. This might be because the higher cognitive ability became more adaptive because of the increased competition among the larger group size.

Second, between around the 1250th and around the 1400th generations (2) in Fig. 4(a), PP and $Gsize$ started to decrease. This means that the linguistic groups diversified because of the extension of the possible range of linguistic group sizes in the previous phase. Also, the variances in both the characteristics of languages and the characteristics of agent abilities, increased temporally. In this way, the linguistic groups diversified in this phase. We think that high communicative elites with low plasticity appeared as the results of these processes. Furtheremore, at the same time, the average of the cognitive linguistic abilities started to decrease. This might be because the lower cognitive ability became more adaptive because of the decreased competition among the smaller group size.

Third, from around the 1400th generation until around the 1500th generation (3) in Fig. 4(a), $NumL$ and $VarCA$, PP decreased and $ComA$ rapidly increased. We think this means that the quick directional selection occurred because high communicative elites with low plasticity appeared and invaded into the whole population very quicky. After that, through a cultural change, the language population was pulled by the users who had survived because of both their high communicative linguistic abilities and small plasticity costs. Then, the average expressiveness in languages caught up with the average communicative linguistic ability in the users, i.e., both the individual population and language population moved outward from the origin, and the overall system state was back to the initial state of a cycle.

As these cyclic coevolution processes repeated, the average expressiveness in languages and the average communicative linguistic abilities in individuals increased. These cyclic coevolution processes seem to be basically the same as the ones found in [5]. Also, these cyclic coevolution processes seem to be similar regardless of the value of the resource finiteness to some extent. In fact, we

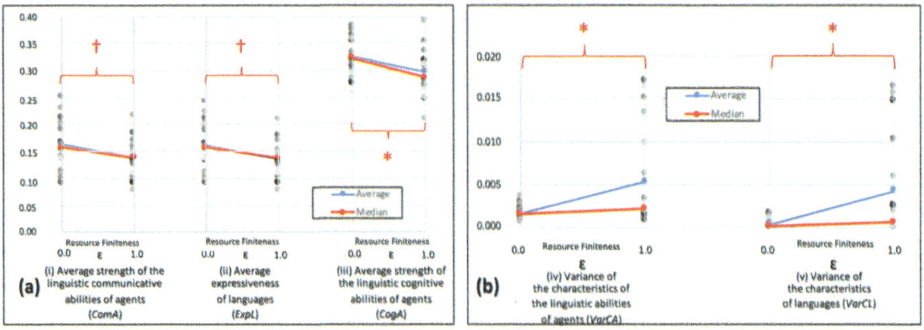

Fig. 3. Averages and medians over 20 trials of averages during the last 500 generations of (i)–(v), when $\epsilon = 0.0$ and $\epsilon = 1.0$.

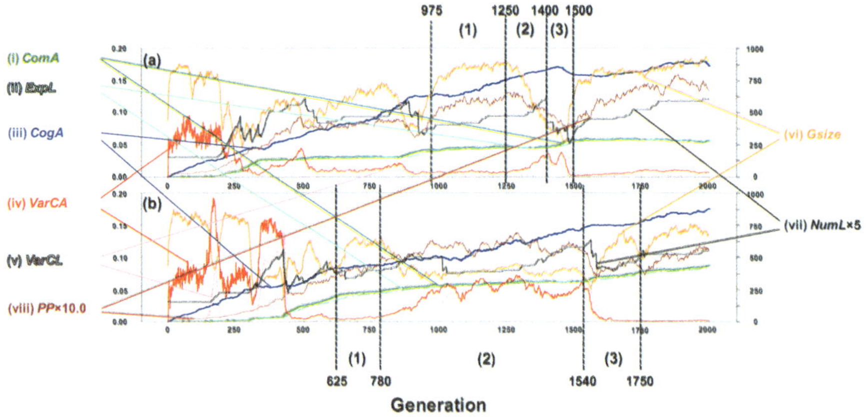

Fig. 4. Time evolution of (i)–(viii) over the first 2000 generations among all the 5000 generations in one trial when $\epsilon = 0.0$ (a) and $\epsilon = 1.0$ (b).

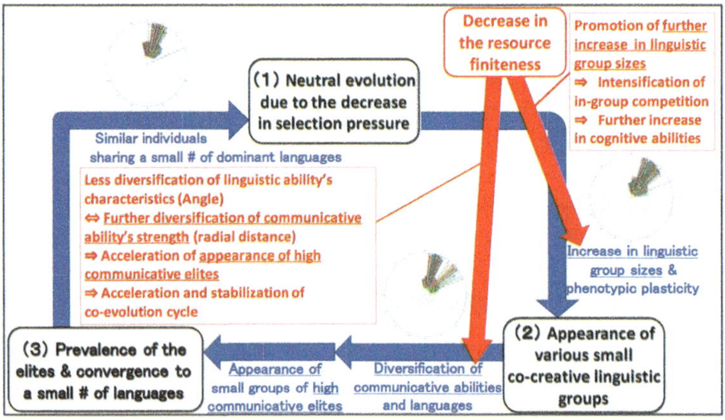

Fig. 5. Cyclic coevolution process of individual linguistic ability and language.

can see a similar cyclic coevolution process from around the 625th generation to around the 1750th generation ((1)–(3)) in Fig. 4(b). However, we could also see that there were differences between when $\epsilon = 0.0$ and when $\epsilon = 1.0$. The cyclic coevolution processes when $\epsilon = 0.0$ were clearer and less fluctuated (more stable) than the ones when $\epsilon = 1.0$. This might be because the decrease in the linguistic diversity (in terms of variance of the characteristics of the linguistic abilities) due to the decrease in the resource finiteness, which is shown in Fig. 4(b), accelerated the cyclic coevolution process of the linguistic abilities for individuals and languages, via the acceleration of the diversification of communicative ability' stlength.

4 Conclusion

We aimed to investigate how a shift from material to informational aspects of shared resources can promote language evolution using a computer simulation. We focused on the two levels of coevolution: that of biological and cultural evolution, and that of two aspects of linguistic abilities. We also extended our model mainly by incorporating the resource finiteness. The experiments showed that a decrease in the resource finiteness could promote the language evolution while decreasing their linguistic diversity in terms of their characteristic variation.

Acknowledgement. This work was supported in part by JSPS/MEXT KAKENHI: JP17H06383 in #4903.

References

1. Smith, K.: How culture and biology interact to shape language and the language faculty. Top. Cogn. Sci. **12**(2), 690–712 (2018). https://doi.org/10.1111/tops.12377
2. Harman, G.: Language, thought, and communication. In: Gunderson, K. (ed), Minnesota Studies in the Philosophy of Science. Language, Mind, and Knowledge, vol. 8. University of Minnesota Press, Minneapolis (1975)
3. Yonenoh, H., Suzuki, R., Arita, T.: Effects of individual and social learning on the evolution of co-creative linguistic communication. Artif. Life Robot. **24**(4), 534–541 (2019). https://doi.org/10.1007/s10015-019-00549-7
4. Bickerton, D., Szathmáry, E.: Confrontational scavenging as a possible source for language and cooperation. BMC Evol. Biol. **11**(261) (2011)
5. Azumagakito, T., Suzuki, R., Arita, T.: Cyclic behavior in gene-culture coevolution mediated by phenotypic plasticity in language. In: The Proceedings of the 12th European Conference on Artificial Life, pp. 617–624 (2013)
6. Byrne, R.W., Whiten, A.: Machiavellian intelligence: social expertise and the evolution of intellect in monkeys, apes, and humans. Behav. Philos. **18**(1), 73–75 (1990)
7. Pearson, K.: On lines and planes of closest fit to systems of points in space. Philos. Mag. A **2**(6), 559–572 (1901)

A System for Generating Audio Influenced by Audience Evaluation Using Interactive Genetic Algorithm

Maho Taniguchi(✉), Kense Todo, Shoya Yasuda, and Masayuki Yamamura

Department of Computer Science, School of Computing,
Tokyo Institute of Technology, Yokohama, Kanagawa, Japan
taniguchi.m.ad@m.titech.ac.jp

Abstract. In generation and selection sound effects and music, it is necessary to search through a large audio database to find the right audio for a scene in an animation or other video clip. However, the sound effects or background music generated by sound director may sometimes the background music selected for certain scenes don't match the context of the scenes in movies or shows very well. Therefore, an approach to generate audio considering listeners' preferences is required. In this work, we suggest a way to generate a suitable audio for a scene involved feedbacks from audience. In particular, we used SpecGAN, which generates a wide variety of audio from latent space, and Interactive GA, which is an optimization algorithm that uses human preferences in evaluation. After SpecGAN generates audio from latent variables, audio is ranked by humans, and the best group of latent variables were crossed over for create the next latent variables. As a result, we succeeded in controlling the direction of generating audio along individual scene.

Keywords: Genetic algorithm · Interactive GA · SpecGAN · Human in the loop · Sound optimization · Evolutional algorithm

1 Introduction

In the generation and selection of music/sound effects, it is necessary to search large audio databases to find an appropriate audio for the scene of animation or other video clips. However, music/sound effects generated by sound directors may sometimes not be suitable for the audience. Therefore, there is a need for an approach that generating audio in accordance with the preferences of the audience.

Recently, several methods using Deep Neural Networks (DNN) have been proposed for research on speech recognition and speech synthesis. WaveNet [5] is a convolutional DNN that predicts audio by autoregressive model. According to Chris Donahue et al., audio generation on WaveNet can take a long time [1]. MuseNet [6] is a DNN to generate a music combined some characteristic of existing composers. While these generation methods can be able to generate musics strictly along the dataset, there is no way to consider human preferences.

K. Yada et al. (Eds.): JSAI 2020, AISC 1357, pp. 208–215, 2021.
https://doi.org/10.1007/978-3-030-73113-7_20

There are researches for producing target audios based on evaluations of multiple people. HUMANGAN [2] is a hostile Generation Adversarial Network (GAN) that incorporates human evaluation into discriminator. It is intended to modeling the human preference distributions. "Sign Sounds Family Generation System using the Interactive Genetic Algorithm" [4] proposed the system that generate some sign sounds by using Interactive Genetic Algorithm (Interactive GA). In both methods, learning and human evaluation are being promoted simultaneously in the optimization process. This is a bottleneck in terms of production efficiency. Therefore, limitation of numbers of evaluation prevents a lot of possible audio from generation.

In this paper, we propose a new architecture that combines Interactive GA, which can be controlled by human, with SpecGAN, which specializes in audio generation, for generating audio that match human preference. SpecGAN is pretrained to learn a variety of sounds in a large latent space. A specific theme was given to the evaluators in advance. An optimization was performed by repeating following two steps; SpecGAN generator generates sounds, and latent values were evolutionally changed as human evaluation in Interactive GA. Sounds before optimization are chaotic, however it is expected that these sounds will be more in according with a specific theme after optimization. As objective evaluation, we confirmed that the sounds can be assigned certain characteristics in terms of the peak level and frequency during optimization. In addition, we could also confirmed that the system can reflect the evaluator's preferences.

2 Proposed Method

In this paper, we propose several methods to reconstruct the target audio using SpecGAN and Interactive GA. Sounds that we prepared SpecGAN of which learning has been finished. Therefore the input vector of SpecGAN generator was optimized using Interactive GA. SpecGAN generator have learned audio data set in advance. To perform Interactive GA, the generator and human evaluation was combined (Fig. 1).

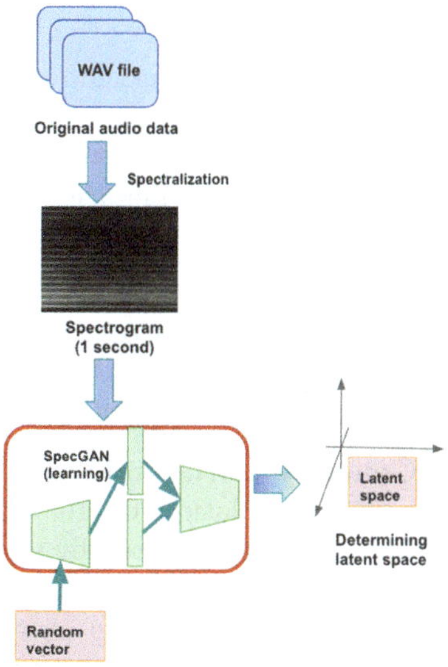

Fig. 1. The flow until determining latent space with SpecGAN.

2.1 SpecGAN

GAN is an unsupervised learning method for mapping low-dimensional latent vector to high-dimensional data, and shows some examples of successful image generation. In SpecGAN, audio data is converted into spectrogram using short-time Fourier transform. The spectrogram is then converted into image data and learned by general DCGAN. The generated image was treated as a spectrogram and reconstructed into audio data by short-time inverse Fourier transform.

In this study, we used the test set from NSynth dataset [3] for training data. Table 1 shows the structure of SpecGAN generator. Table 2 shows details of the NSynth dataset.

Table 1. SpecGAN generator structure.

Operation	Output shape
Input $z \sim$ Uniform [−1, 1]	(n, 100)
Dense	(n, 256d)
Reshape	(n, 4, 4, 16d)
ReLU	(n, 4, 4, 16d)
Conv2D	(n, 8, 8, 8d)
ReLU	(n, 8, 8, 8d)
Conv2D	(n, 16, 16, 4d)
ReLU	(n, 16, 16, 4d)
Conv2D	(n, 32, 32, 2d)
ReLU	(n, 32, 32, 2d)
Conv2D	(n, 64, 64, d)
ReLU	(n, 64, 64, d)
Conv2D	(n, 128, 128, c)
Tanh	(n, 128, 128, c)

Table 2. Detail of NSynth dataset (test set).

Category	The number of data
Bass	843
Brass	269
Flute	180
Guitar	652
Keyboard	766
Mallet	202
Organ	502
Reed	235
String	306

2.2 Interactive GA

Interactive GA is a method based on Genetic Algorithm(GA), of which computational evaluation part is alternated by human evaluation. GA is a method of obtaining an optimal solution by repeatedly performing operations of selection, crossover, and mutation. In this study, only selection and crossover were incremented. Table 3 shows the parameters used in Interactive GA.

Genetic Manipulation. The optimization method using Interactive GA is shown below. Figure 2 shows a flowchart.

I. Generate initial individuals
The 100 population of 100-dimensional latent vector is generated by uniform random numbers in the range of $[-1, 1]$. After the generation, the average of the Euclidean distance between vectors is calculated and the lowest one was replaced with a newly generated vector to increase

Table 3. Parameters for Interactive GA.

Category	The number of data
Population	100
Parent	4
Child	16
Generations	200
Crossover	REX
Extraction	Unrestored extraction
Selection	Tournament selection

the diversity of the initial population if the average of distance was increased. We continued this operation until it is no longer updated 1000 times in a row.

II. Extract parent individuals
Four individuals are randomly extracted from current population as a parent group.

III. Generate child individuals
Using a parent group, 16 child individuals are generated by REX crossover. The centroid of the parent individuals $x_1, \cdots, x_n$ was represented by x_g, and a child individual is generated according to the following equation.

$$x_c = x_g + \sum_{i=1}^{n} \xi_i (x_i - x_g) \ , \quad \xi_i \sim \varphi(0, \alpha\sigma_\xi^2) \tag{1}$$

where $\alpha = 1.1$, which is based on the results of prior experiments. $\varphi(0, \sigma_\xi^2)$ is normal distribution with mean 0 variance σ_ξ^2, and σ_ξ^2 has impose conditions.

$$\sigma_\xi^2 = \frac{2}{n} \tag{2}$$

IV. Evaluation, selection
A human evaluator select the top 4 individuals from the 16 child individuals by tournament selection. In tournament selection, always two individuals were presented at once, to perform a pair comparison to select "better" one.

V. Replace
The best four individuals selected from the offspring were replaced with the four individuals selected as parent individuals. After replacing it, go back to 2 and repeat. The process of 2 to 5 was defined as one generation.

Target Sound and Subject Restriction. In this paper, we prepared two target audio to examine the validity of this method. One target is a "plosive audio" and the other is a "healing audio". In order to check whether the final audio is different, we chose a "healing audio" that give a positive emotion and a "plosive audio" that gives a negative emotion such as discomfort and fear to confirm the final audio. Subjects were eight human. Each subject was given a purpose and evaluated. Each generated audio was evaluated one by one, and there was no duplication in the audio data evaluated.

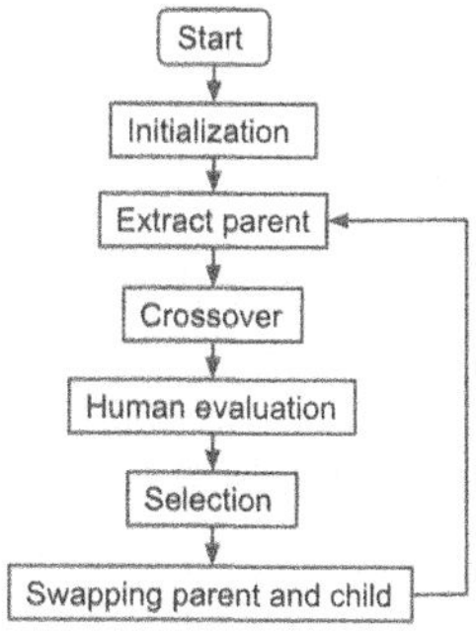

Fig. 2. Flowchart of interactive GA.

3 Evaluation

3.1 Quantitative Evaluation of Interactive GA

We qualitatively evaluate the audio generated from each 100 individuals of each target selected by Interactive GA and the initial individuals. We have constructed two indicators, peek level and the frequency of peek level were compared. We also compared between several generations (0, 50, 100, 150, 200) of plosive and healing sounds to see how they evolved.

3.2 Human Evaluation

A human evaluation was conducted to ensure that the interactive GA optimizations matched the perceptions of those who participated in the optimizations. People who had not participated in the optimization were also asked to evaluate the sound in order to confirm that the sound matches their perceptions. Four people who participated in the optimization process and six people who did not participate in the optimization process rated the sound. We asked each person to listen to 21 sounds (seven optimized, seven un-optimized, and seven ready-made sounds [8]) for each theme. They chose to evaluate the fitness for each theme four levels(suitable, permissible, not permissible, wrong).

4 Results

4.1 Interactive GA Evaluation Results

(a) and (b) of Fig. 3 show the peak level and frequency at the peak each 100 individuals of "initial individuals", "plosive audio" and "healing audio". (c) and (e) show the peak level of each generation for plosive and healing sounds, and red(d) and (f) show the frequency of each generation.

Tables 4 and 5 show the means, standard deviations, and medians of the peak level and frequencies of plosive and healing sounds for several generations (0, 50, 100, 150, and 200) for each of the initial 100 individuals. Comparing the peak level at the peak, the "plosive audio" was higher than the "healing audio" in both the average and the median. Therefore the plosive sound tends to be loud. Comparing the frequency at the peak, the "healing audio" tended to have a higher frequency than the "plosive audio".

When "plosive audio" and "healing audio" were applied to Welch's t test for both indexes, the result was $pvalue < 0.01$, indicating a significant difference at the significance level 1%. As a result, these two targets are producing different audio quantitatively.

When looking at the volume of rupture sounds across generations, there is a downward trend in the generations near the initial generation and an upward trend in the middle to final generation. This is likely due to the fact that middling sounds, such as those that are just too loud, were eliminated in the initial optimizations and the optimization to plosive sounds in the mid to final generation. The other three showed a trend of convergence, with the box beard getting smaller as the generation progressed.

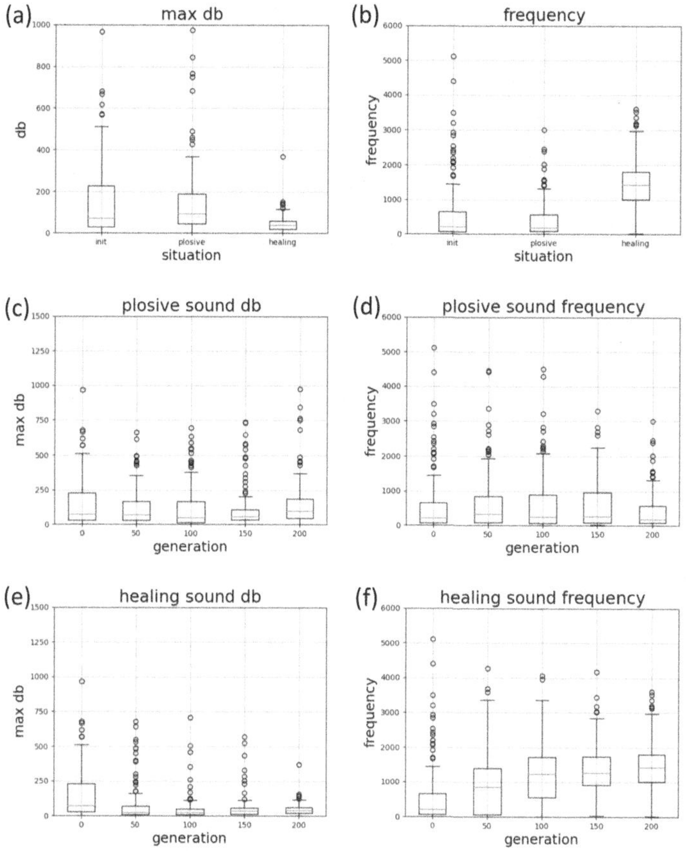

Fig. 3. Peak levels (a) and frequencies at the peak (b) of initial, optimized plosive, optimized healing sounds. Peak levels and frequencies along generation of plosive (c–d) and healing sound (e–f). Each box consists of population size of 100.

Table 4. Value at maximum volume. (n = 100)

Experiment	Average	Standard deviation	Median
Initial	153.02	184.72	73.05
Plosive (generation 50)	126.22	147.94	68.15
Plosive (generation 100)	124.06	167.67	47.73
Plosive (generation 150)	122.40	166.87	57.92
Plosive (generation 200)	159.88	186.47	96.23
Heling (generation 50)	88.78	149.76	23.40
Heling (generation 100)	54.45	104.53	24.26
Heling (generation 150)	59.99	96.44	36.88
Heling (generation 200)	47.59	45.93	39.94

Table 5. Frequency at maximum volume. (n = 100)

Experiment	Average	Standard deviation	Median
Initial	658.98	1005.15	216.54
Plosive (generation 50)	717.11	938.84	325.79
Plosive (generation 100)	790.32	1218.24	238.68
Plosive (generation 150)	624.29	764.15	262.30
Plosive (generation 200)	478.42	600.62	188.48
Heling (generation 50)	999.54	1005.11	853.35
Heling (generation 100)	1249.00	909.75	1230.81
Heling (generation 150)	1301.10	823.97	1263.29
Heling (generation 200)	1505.57	916.42	1437.01

4.2 Human Evaluation

Table 6 shows human evaluations by those who participated in the interactive GA optimization. Table 7 shows evaluations by those who did not participate in the optimization.

The evaluations of those involved in the optimization were more than twice as many as the first, with a score of 200 for "suitable+permissible", so the sound is more suited to the evaluator's perception. However, it has less than half the number of ratings compared to the ready-made sound and needs to be optimized more efficiently.

In the case of the plosive sound, There was no difference between the first and 200 generations. In the case of the healing sound, the number of people who felt it was not a wrong sound was about half of those who felt it was not a wrong sound, and even those who were not involved in the optimization process felt it was closer to the healing sound.

Table 6. Evaluation of plosive and healing sounds by those PERFORMED optimization task. Each four evaluators votes seven variations of ready-made, initial and optimized sounds.

	Plosive			Healing		
	Ready-made	200 gen	First gen	Ready-made	200 gen	First gen
Suitable	23	3	2	19	4	0
Permissible	5	8	4	5	13	4
Not permissible	0	12	9	4	10	5
Wrong	0	5	13	0	1	19

Table 7. Evaluation of plosive and healing sounds by those NOT performed optimization task. Each six evaluators votes seven variations of ready-made, initial and optimized sounds.

	Plosive			Healing		
	Ready-made	200 gen	First gen	Ready-made	200 gen	First gen
Suitable	22	0	1	5	0	0
Permissible	9	2	4	16	5	0
Not permissible	7	7	8	10	19	9
Wrong	4	33	29	11	18	33

5 Conclusion

In this research, we constructed a system that generates sound according to the preference of audience by combining SpecGAN and Interactive GA. From experiment results, the referent characteristics appeared for each target scene in terms of quantify.

During evaluation, we found that the related characteristics were appeared in terms of peak level and frequency. In addition, as a subjective evaluation, we confirmed that the optimal sound is reflected by the evaluator's preferences. Therefore, it was demonstrated that the system was able to generate sounds in accordance with a specific theme. Interestingly, different people have different preferences, even if with the same theme. This suggests that Interactive GA can control generating general sounds with a large number of evaluators, or generating specific sounds with a small group. In the future, this research can be extended for SpecGAN to synthesize longer sounds to handle various sound targets.

References

1. Donahue, C., McAuley, J., Puckette, M.: Adversarial audio synthesis. arXiv preprint arXiv:1802.04208 (2018)
2. Fujii, K., Saito, Y., Takamichi, S., Baba, Y., Saruwatari, H.: HumanGAN: generative adversarial network with human-based discriminator and its evaluation in speech perception modeling. In: ICASSP 2020–2020 IEEE International Conference on Acoustics, Speech and Signal Processing (ICASSP), pp. 6239–6243. IEEE, May 2020
3. Engel, J., et al.: The NSynth dataset (2017). https://magenta.tensorflow.org/datasets/nsynth
4. Hiroko, M.M., Maiko, O., Sugahara, Wake, S.H., Morishita, C., Hiroyasu, T.: Sign sounds family design system using an interactive genetic algorithm. J. Hum. Interface Soc.: Hum. Interface **10**(2), 243–252 (2008)
5. Oord, A.V.D., et al.: WaveNet: a generative model for raw audio. arXiv preprint arXiv:1609.03499 (2016)
6. Payne, C.: MuseNet. OpenAI Blog (2019)
7. Kobayashi, S.: Trans. Jpn. Soc. Artif. Intell.: AI **24**(1), 147–162 (2009)
8. Sound effect Lab. https://soundeffect-lab.info

Strategy and Spatial Representations for Improving Navigation Ability Without Vision Using Sensory Substitution Devices

Kouta Akimoto(✉), Ayako Fukawa, and Ikuko Eguchi Yairi

Sophia University, Tokyo, Japan
akimoto@yairilab.net

Abstract. Accurate and quick navigation poses a major challenge for the visually impaired and those who have been hit by power outages, smoke and fog. SSDs are expected as a technical approach to this problem. The improvement of navigation by SSD and the knowledge of neuroscience and cognitive psychology are not so linked. In this study, we examine sensory information and strategies related to improving navigation by reviewing existing SSDs from the perspective of neuroscience and using the triangle completion task that has been used for a long time in cognitive psychology. From reviewing SSDs, it was found that depth information, direction information, and object information provide the information necessary for constructing spatial representations instead of vision. We found out that a strategy of recalling triangles using Labeled graphs is effective in the long term for improving navigation. Directional information encourages the construction of labeled graphs, and feedback on the direction of the navigation goal makes it accurate via the acquisition of metaknowledge and improvement of subjective scale. Our research would make a great contribution to the development of SSDs for navigation.

Keywords: Navigation · Sensory substitution device · Neuroscience · Psychology · Triangle completion task · Vibrotactile feedback · Spatial representation

1 Introduction

Navigation is an essential task in our lives. Vision provides the most reliable information about navigation. The challenge is accurate navigation assistance to quickly move to a safe place for an estimated 252 million VIs [1] and those who have been hit by power outages, smoke and fog. Sensory substitution devices (SSD) are expected as a technical approach to this problem. SSD is a non-invasive device that converts visual information into tactile and auditory information. Few SSDs have been developed based on the knowledge of neuroscience and cognitive psychology. The purpose of this study is to improve the ability to navigate without vision in a small space, find the external information necessary for improving the ability, verify it from the knowledge of neuroscience and cognitive psychology, and bring new knowledge to human computer interaction.

K. Yada et al. (Eds.): JSAI 2020, AISC 1357, pp. 216–223, 2021.
https://doi.org/10.1007/978-3-030-73113-7_21

2 Background

2.1 Navigation Without Vision in Small Spaces

Navigation strategies can be broadly divided into the following two categories: (1) Local navigation (2) Wayfinding [2]. Local navigation is a strategy used when the destination can be observed directly from the current position, and can be used without thinking of a map in the head. It can be used without thinking. Wayfinding is a strategy used when the destination cannot be observed directly from the current position, and it is necessary to think of a course or map in mind and refer to them. Without vision, the destination cannot be observed directly, so wayfinding is necessarily used. Of the wayfinding, recognition-triggered response and topological navigation are strategies that repeat local navigation from landmark to landmark [3, 4], so they cannot be used without vision. As a result, the survey strategy must be taken in the route search and a map will be drawn in the head. In small spaces such as indoors, ends of the passage, obstacles, and corners of the room would be landmarks.

2.2 Spatial Representations

Spatial representation is the map that you envision in your head when you take a research strategy. We form spatial representation by investigating the environment [5].For example, when you visit a new city, we will stroll through the city and gradually learn about hotels, landmarks, streets, and intersections. And when you go to a supermarket, you can get to the location from spatial relationships between other landmarks without knowing the exact location and direction. Figure 1 shows three forms of spatial representation [6]. Graph knowledge (Fig. 1A) expresses the adjacency between places. Survey knowledge (Fig. 1B) expresses the positional relationship between places by coordinate system on an Euclidean scale. Survey knowledge is constructed by path integration during spatial learning. Labeled graphs represent graph knowledge between locations and approximate distances and angles. An individual's spatial representation depends on learning time, individual decision-making, and required tasks.

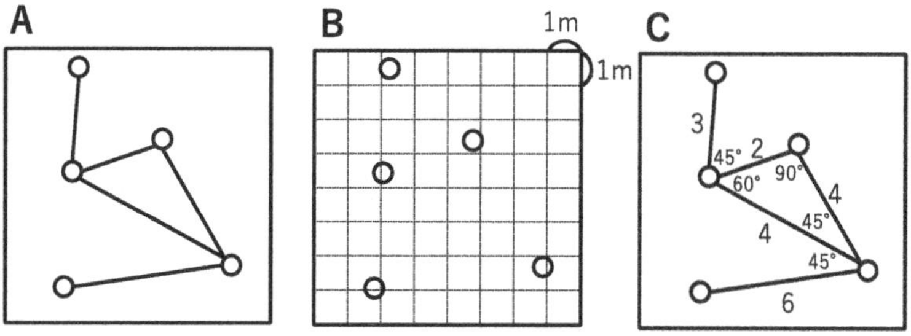

Fig. 1. Three forms of spatial representation.

2.3 Measurement of Navigation Ability

Navigation is a multimodal process that involves perception, short-term memory, long-term memory, attention to landmarks, and decision making [7]. Various models of navigation system and neural spatial representations have been proposed and verified. Path integration is a model that estimates the travel distance and turn angle by integrating the information provided from the vestibular, kinesthetic sense, proprioception, and optic flow, and updates the current position as we move [8]. In this study, we focus on route integration ability as a navigation ability without using vision.

3 Review

In this study, we will collect research on SSDs used to improve navigation ability, identify important sensory information from the viewpoint of neuroscience, and describe its relationship with the formation of spatial expressions. The purpose of the review is to confirm the usefulness of the devices by linking the disparate findings of neuroscience and cognitive psychology in navigation research.

3.1 Method

We searched for the mainly keyword "navigation" and "sensory substitution device" using the bibliographic database PubMed and Google Scholar. We investigated the 9 papers that showed improvement in navigation-related ability.

3.2 Depth Information

Depth information allows us to detect obstacles in front of us. Depth information is also used when bats and dolphins perform echolocation. Devices that present depth information include TVSS [9], MeloSee [10], EyeCane [11], SoV [12], vOICe and BrainPort [13]. In the maze task using EyeCane, it was confirmed that the correct path of the maze could be drawn more accurately. The depth information provides the traversable path, room shape, and distance to obstacles. Since the walls and corners of the room are fixed in the environment, they serve as landmarks. Paths and landmarks of the spatial representations are constructed with depth information.

In neuroscience, border cells have been found in the hippocampus of rats [14]. Border cells fire at locations near boundaries, such as near walls in the environment [15]. The perception of depth information is also important from the existence of border cells.

3.3 Direction Information

Direction information allows us to easily orient. Migratory birds and sea turtles obtain direction information from their magnetic sensations to return to their nests. Devices that present direction information included belt-type tactile devices [16, 17] and headphone-type auditory devices [18]. Kaspar showed in a questionnaire to belt wearers that their navigational abilities were subjectively improved [17]. Direction information provides the exact angles between paths needed for the Labeled graph. It also enables reorientation when our direction is lost.

In the field of neuroscience, head direction cells have been found in rodents [19]. One head direction cell fires when facing a specific direction. Perception of direction information is also important from the presence of head direction cells.

3.4 Object Information

Object information allows landmarks to be identified. Object information is presented by assigning different sounds to each object [20] or brightness [13]. Object information identifies landmarks and recalls specific scenes. Scene recognition makes it possible to quickly grasp the current location by referring to the spatial representations.

In the field of neuroscience, spatial view cells have been found in primates [21]. A spatial view cell is a cell that fires when a specific landscape enters the field of view. Spatial view cell also helps to remember place and its context, such as the distance to other places and episodes [22]. Object information is also important from the existence of spatial view cells.

4 Experiment

In this study, we investigate the effect of vibration and pulling on short-term and long-term path integration ability in a small space using the Triangle Completion Task (TCT). TCT is one of the homing tasks in which a participant moves while being guided on two segments of a triangle, and responses a starting point vector (Homing Vector, HV). In this study, the path integration ability is measured using the HV angle error. By feeding back the angle error of HV, we investigate the change of path integration ability before learning, short term (during the experiment) and middle term (one week later). The purpose of this study is to identify the relationship between each participant's strategy and spatial expression and to verify the effect of external information according to the navigation goal from data obtained by experiments.

4.1 Methods

Participants: Participants were university and graduate students (nine males and one female), average age 21.7 years (SD 0.7), and healthy individuals with normal hearing and vision. Participants were randomly divided into a vibration & pulling group provided feedback using vibration and traction and a vibration group provided feedback only using vibration.

Materials: The experiment was performed in an 11m x 8m room. The vertex of the path was marked on the floor with color tape. Participants put 5cm x 5cm gauze on both eyes and put on blindfolds to block vision. In addition, we used a smartphone and earphones that recorded white noise to block hearing. To measure the direction of the origin, we used a 2m tension rod and a digital goniometer. A chair with casters was used to guide the participants. A head-mounted vibration device (hereinafter called a vibration device) has been developed that can present information in 16 directions to indicate the origin direction. The vibration device measures the participant's rotation, and can always present the origin direction.

Tasks: There were two types of tasks: a direction judgment task and a learning task. The 12 paths were performed in random order for each participant. Paths composed of a combination of two segments (3, 6m) and three angles (60, 90, 120°).

The homing task is a task in which the participant places the tension rod in the direction they think to be the origin after moving. After that, the vibration & pulling group was moved by the experimenter to the origin by the shortest distance. The Vibration group was pulled by the experimenter to the origin on a random winding route. It took more than 20 s to move to the origin. One direction determination task lasted about 20 min.

In the learning task, as in the case of the homing task, the participants placed the tension rod in the direction they think to be the origin after moving. After that, the experimenter mounted the vibration device on the participant and presented the direction of the origin by vibration based on the measured angle. Feedback was given until the participants turned in the right direction. The movement to the origin was performed in the same way as the homing task. Each learning task lasted about 40 min.

Procedure: All experiments were performed by one experimenter. At the beginning of the experiment, the experimenter described the paths as "there are 12 types of paths consisting of linear movement and one turn", and "all paths are different". The experimenter explained to the participants about the tasks. Next, the participants practiced slowly placing the tension rod in the direction in which they were facing. Participants also practiced turning the body in the direction presented by wearing a vibration device. After that, participants walked the path for practice with the guidance of the experimenter and practiced the homing task. The experimenter shouted to the participants when they started walking, turned around, and stopped.

The homing task was performed three times, once before and after the learning task, and once after a week. Participants took a 5–10 min break between each task. In each task, the participant who blocked vision and hearing was guided to the origin by the experimenter. The total experiment time was about 2 h. After the experiment, the experimenter interviewed participants about the method used to determine the origin.

4.2 Results

The participants assigned to the vibration & pulling group were S1, 2, 3, 9, 10 and the participants assigned to the vibration group were S4, 5, 6, 7, 8. TCT errors are caused by the following three factors. (1) Encoding path distances and turn angles into spatial representations (2) Integrating information for determining HV (3) Performing rotation and walking [23]. P-values with multiple tests were corrected according to the Bonferroni method. An asterisk was used to indicate statistical significance, as follows: $p < .05(*)$. The numerical data were analyzed with Excel.

Regarding the relative error average, there was a significant difference in S1, 2, 4, 5 between before leaning and short-term, and S4, 8, 10 between before learning and mid-term. For all participants, if the short-term relative error average was negative, it increased, and if it was positive, it decreased. The same tendency was observed for mid-term except for S2.

Table 1 shows the strategies and absolute error average for participant. In S1, 3, 5, 6, 8, 9, 10, HV was calculated geometrically based on the mental image of the triangle derived from the number of steps and the sense of rotation. S2, 7 judged HV by intuition. S4 maps the end points and origin to the mental image of the grid and determines the HV on the grid. Regarding the absolute error average, there was a significant difference in S1 between before learning and short-term. In addition, it decreased except for S4, 10. There was a significant difference in S1 and S3 between before learning and mid-term. In addition, it decreased except for S8.

Table 1. Strategy and Absolute error average for each participant.

Participants	Strategy	Before	Short-term	Mid-term
S1	Thinking of a triangle	37.9	21.4*	19.6*
S3	Thinking of a triangle	29.5	18.1	10.2*
S5	Thinking of a triangle	18.4	16.4	11.0
S6	Thinking of a triangle	19.7	14.5	19.0
S8	Thinking of a triangle	14.8	14.7	31.3
S9	Thinking of a triangle	12.5	12.4	13.1
S10	Thinking of a triangle	24.4	28.0	16.0
S2	Intuition	13.3	10.1	12.0
S7	Intuition	17.2	12.5	14.1
S4	Mapping nodes	11.2	15.2	9.2

5 Discussion

For most subjects, the relative error fluctuated in the opposite direction, positive and negative, compared to before learning. S5 was conscious of correcting the bias in short-term and later tasks. Therefore, participants will have gained meta-knowledge about rotational bias through vibration feedback. Given the decreasing tendency of the absolute error, meta-knowledge would be refined by repeated learning.

S2 and 7 were judged by intuition. The absolute error did not decrease from short-term to mid-term. Therefore, they would not have constructed a spatial representation and the learning effect would be the acquisition of only meta-knowledge. S4 remembers a grid in his head. The absolute error decreased from the short term to the middle term. So S4 would have preferred Survey knowledge and the learning effect would be meta-knowledge and the Survey knowledge of paths used in this experiment. S1, 3, 5, 6, 8, 9, 10 came to mind the image of a triangle. In addition, there were large individual differences in the perception of the rotation angle. So they would have built the Labeled graph and the learning effect would be meta-knowledge and improving of subjective scale accuracy. A strategy that is expected to have long-term learning effects would be to build the Labeled graph.

As a result of the review, depth, direction and object information contribute to the formation of spatial representations instead of vision. In particular, the direction information gives the information necessary for constructing a Labeled graph. Therefore, the presentation of direction information and the feedback indicating the destination of navigation promote the construction and improvement of the Labeled graph, and a long-term effect can be obtained.

6 Conclusion

In this study, we conducted the review and the experiment using the TCT in order to improve the navigation ability without vision. In order to improve navigation ability, it is first necessary to construct a labeled graph, and directional information is useful for that. And feedback in the direction of the destination is effective to make the labeled graph accurate. This is because it promotes the acquisition of meta-knowledge about navigation and the improvement of the accuracy of subjective scales.

These findings should be used in the future in the development of devices that contribute to the discovery of neurons involved in navigation and in the development of navigation support devices and UI designs that take into account personal preferred navigation methods and spatial representations.

Acknowledgement. The authors would like to thank participants for cooperating with experiments. This research was supported by Grants-in-Aid for Scientific Research B (17H01946).

References

1. Bourne, R.R.A., Flaxman, S.R., Braithwaite, T., et al.: Magnitude, temporal trends, and projections of the global prevalence of blindness and distance and near vision impairment: a systematic review and meta-analysis. Lancet Glob. Heal. **5**(9), e888–e897 (2017)
2. Trullier, O., Wiener, S.I., Berthoz, A., Meyer, J.A.: Biologically based artificial navigation systems: review and prospects. Prog. Neurobiol. **51**(5), 483–544 (1997)
3. Franz, M.O., Mallot, H.A.: Biomimetic robot navigation. Rob. Auton. Syst. **30**(1), 133–153 (2000)
4. Eichenbaum, H.: The role of the hippocampus in navigation is memory. J. Neurophysiol. **117**(4), 1785–1796 (2017)
5. Wang, R.F.: Building a cognitive map by assembling multiple path integration systems. Psychon. Bull. Rev. **23**(3), 692–702 (2015). https://doi.org/10.3758/s13423-015-0952-y
6. Chrastil, E.R., Warren, W.H.: From cognitive maps to cognitive graphs. PLoS ONE **9**(11), e112544–e112544 (2014)
7. Ekstrom, A.D., Huffman, D.J., Starrett, M.: Interacting networks of brain regions underlie human spatial navigation: a review and novel synthesis of the literature. J. Neurophysiol. **118**(6), 3328–3344 (2017)

8. Loomis, J.M., Klatzky, R.L., Golledge, R.G., Philbeck, J.W.: Human navigation by path integration. Wayfinding Behav. Cogn. Mapp. Spatial Process. 125–151 (1999)
9. Segond, H., Weiss, D., Sampaio, E.: Human spatial navigation via a visuo-tactile sensory substitution system. Perception **34**(10), 1231–1249 (2005)
10. Stoll, C., Palluel-Germain, R., Fristot, V., Pellerin, D., Alleysson, D., Graff, C.: Navigating from a depth image converted into sound. Appl. Bionics Biomech. **2015**, 1–9 (2015). https://doi.org/10.1155/2015/543492
11. Chebat, D.R., Maidenbaum, S., Amedi, A.: Navigation using sensory substitution in real and virtual mazes. PLoS ONE **10**(6), 1–18 (2015)
12. Hoffmann, R., Spagnol, S., Kristjánsson, Á., Unnthorsson, R.: Evaluation of an audio-haptic sensory substitution device for enhancing spatial awareness for the visually impaired. Optom. Vis. Sci. **95**(9), 757–765 (2018)
13. Jicol, C., Lloyd-Esenkaya, T., Proulx, M.J., et al.: Efficiency of sensory substitution devices alone and in combination with self-motion for spatial navigation in sighted and visually impaired. Front. Psychol. **11**, 1443 (2020)
14. O'Keefe, J.: Place units in the hippocampus of the freely moving rat. Exp. Neurol. **51**(1), 78–109 (1976)
15. Solstad, T., Boccara, C.N., Kropff, E., et al.: Representation of geometric borders in the entorhinal cortex. Science **322**(5909), 1865–1868 (2008)
16. Kärcher, S.M., Fenzlaff, S., Hartmann, D., et al.: Sensory augmentation for the blind. Front. Hum. Neurosci. **6**(MARCH 2012), 1–15 (2012)
17. Kaspar, K., König, S., Schwandt, J., König, P.: The experience of new sensorimotor contingencies by sensory augmentation. Conscious Cogn. **28**(1), 47–63 (2014)
18. Schumann, F., O'Regan, J.K.: Sensory augmentation: integration of an auditory compass signal into human perception of space. Sci. Rep. **7**(January) (2017)
19. Taube, J.S.: The head direction signal: origins and sensory-motor integration. Annu. Rev. Neurosci. **30**(1), 181–207 (2007)
20. Massiceti, D., Hicks, S.L., van Rheede, J.J.: Stereosonic vision: exploring visual-to-auditory sensory substitution mappings in an immersive virtual reality navigation paradigm
21. Rolls, E.T.: Spatial view cells and the representation of place in the primate hippocampus. Hippocampus **9**(4), 467–480 (1999)
22. Rolls, E.T., Xiang, J.-Z.: Spatial view cells in the primate hippocampus and memory recall. Rev. Neurosci. **17**(1–2), 175–200 (2006). https://doi.org/10.1515/REVNEURO.2006.17.1-2.175
23. Fujita, N., Klatzky, R.L., Loomis, J.M., Golledge, R.G.: The encoding-error model of pathway completion without vision. Geogr. Anal. **25**(4), 295–314 (1993)

A Politeness Control Method for Conversational Agents Considering Social Relationships with Users

Tomoki Miyamoto[1](✉), Daisuke Katagami[1], and Mayumi Usami[2]

[1] Graduate School of Tokyo Polytechnic University, Atsugi-shi 243-0297, Japan
d1985001@st.t-kougei.ac.jp

[2] National Institute for Japanese Language and Linguistics, Tachikawa-shi 190-0014, Japan

Abstract. In this paper, we propose an autonomous linguistic consideration (politeness) control method for conversational agents. As an approach to realize the proposed method, we examined input information based on politeness theory in sociolinguistics that is considered to be effective for agents to control politeness. Specifically, we adapted a questionnaire study to evaluate the politeness and comfortableness of utterances in different situations of requests. The participants evaluated stimulus utterances in terms of three social factors on the politeness theory: social distance/social status difference/conversation scene. Based on the dataset collected in this questionnaire study, a regression model was constructed to predict the politeness and comfortableness rated by the participants of the experiment, adapted three social factors and politeness of utterances as explanatory variables. As a result of this questionnaire study and multiple regression analysis, a valid model of politeness was obtained. The results contribute as a model to select appropriate politeness of utterance according to three social factors on the politeness theory with the user in the proposed method.

Keywords: Conversational agents · Politeness theory · Social relationships

1 Introduction

With the development of technologies related to artificial intelligence and natural language processing, the existence of conversational agents that can interact with humans is becoming more and more familiar. Also, it is expected that a conversational agent with a high level of conversation ability that can function sufficiently as a human conversation partner will be realized. However, research and development of an advanced conversational agent that can express the consideration like a human according to the relationship with the conversation partner by language are still immature.

In general human conversation, humans use linguistic considerations in consideration of the social distance and social power difference with the other party. The methodology of linguistic considerations in the human-to-human conversation has been studied in the fields of pragmatics and sociolinguistics as politeness theory [1, 2]. In contrast, research on Human-Agent Interaction, which focuses on the relationship between humans and

K. Yada et al. (Eds.): JSAI 2020, AISC 1357, pp. 224–231, 2021.
https://doi.org/10.1007/978-3-030-73113-7_22

artifacts, points out the importance of interaction design that allows humans to feel agency to agents (e.g., [3, 4]). Especially in [3], it is reported that the conversational agent uses the politeness strategies defined in [1] is effective to improve the humanity of the agent. Therefore, consideration of politeness by the conversational agents is effective in improving the agency of the conversational agents. It also contributes to the realization of a conversational agent as a human conversational partner.

Therefore, in this paper, we propose a method in which a conversational agent autonomously controls linguistic consideration (politeness). As an approach to realize the proposed method, based on the politeness theory in sociolinguistics, we investigate input information that is considered effective for agents to control politeness. We also discuss how the conversational agent controls politeness at the discourse level based on previous research [1, 2]. We further examined how politeness control is performed in relations between humans based on three factors: social distance/difference in social power /conversation scene (culture), so we report the result. As related works in this paper, there are research for estimating the politeness of question sentences [5] and research for guiding users to conversation groups constructed by multiple virtual agents using politeness strategy [6]. In reference [5], politeness was estimated by machine learning using a text data set labeled with politeness as training data. However, their study only targets situations in which speakers are on an equal footing, and does not verify differences in the evaluation of politeness due to changes in social relationships. Reference [6] focuses on multi-person conversation, so the approach is different from our paper.

2 Social Factors on Human Conversation

According to [1], in a conversation between Speaker (S) and Hearer (H), S tries to keep H's face (i.e., desire in interpersonal relationships) as much as possible. The face is determined by the social distance between S and H (D), the difference in power (P), and the sum (Wx) of the degree of burden exerted by the action (x) in the culture (Rx). S estimates H's face in the conversation. And S uses politeness according to the size of Wx. Therefore, Eq. (1) [1] holds.

$$Wx = D\,(S,\ H) + P\,(H,\ S) + Rx \tag{1}$$

In Eq. (1), $D\,(S, H)$ represents the social distance between S and H, and $P\,(H, S)$ represents the amount of social power that H has over S. If S estimates Wx appropriately, the conversation with H proceeds smoothly. On the other hand, there is a risk that S may incorrectly estimate Wx and threaten H's face. However, in Eq. (1) [1], no threshold value is defined for determining whether the estimate of Wx by S is appropriate. Therefore, it is difficult to explain the error between the face of H and the Wx estimated by S.

For this problem, [2] defined the error between Se and He (i.e., Discrepancy in estimations: $De = Se - He$) for the estimation of Wx by S (Se) and the estimation of Wx by H (He). According to [2], De has an allowable error ($\pm\,\alpha$) between S and H. Therefore, when Eq. (2) holds, both are comfortable in the conversation between S and H.

$$|De| < |\alpha| \tag{2}$$

Equation (2) not only indicates that there is an acceptable margin of error in the estimation of S and H but also when S behaves too polite to H expresses that there is a risk of being greatly threatened the H's face (i.e., $De > \alpha$).

3 Politeness Control Method for Conversational Agent

3.1 Social Factors as Input Information

(1) **Social distance (*D*)**
This changes according to the degree of intimacy between agent and user and the emotion of the user. By applying the method of the previous research [5], it is possible to quantitatively express intimacy by estimating the politeness strategy included in user utterances. There are also APIs (e.g., Face API [7], ST Emotion SDK [8]) for estimating emotions from facial expressions and prosody, and D can be estimated by using them.

(2) **Social power (*P*)**
In human-to-human conversation, it is determined by the age of S and H and the difference in their positions in the organization. In addition, in the conversation between the agent and the user, this may vary depending on the P of the agent assumed by the user. Specific factors include the appearance of the agent/the role of the agent/the user's perception of the artifact. It is thought that the age of the agent assumed by the user fluctuates depending on the appearance of the agent. Assuming that the user considers the age of the agent as a social factor in the same way as human interaction, the age of the user is one of the user input information. Gender can also affect D and P as social factors. The age and gender of the user can be sensed due to the widespread use of Face API [7] that allows highly accurate estimation. On the other hand, an agent is an artificial object, so the agent actually has no concept of gender or age and is set by the developer from the top down. For this reason, depending on the user, the case where P is considered as $user > agent$ irrespective of the appearance or role of the agent may sufficiently occur. For this reason, it is desirable to obtain the user's recognition of artifacts before the conversation with the agent starts. As a concrete method, it is effective to construct a model that can collect questionnaires and collect large-scale data on the relationship between the user's social attributes and recognition of artifacts in advance and estimate them in conversation.

(3) **Degree of burden in the culture (*Rx*)**
The culture of the conversation between the agent and the user is determined by the role of the agent and the language that the user has as the background. For example, for a non-task-oriented conversational agent whose purpose is to chat, casual behavior may be accepted by the user. On the other hand, agents that support tasks as user secretaries are required to have formal behavior. It is also desirable to apply a conversation strategy based on knowledge of pragmatics and sociolinguistics, depending on the language in which the user is the native language. Therefore, this factor is the information that the agent should have as prior knowledge.

3.2 Expression of Conversational State

We believe that by expressing the time series continuously, it is possible to explain politeness control that considers not only discussions divided by one utterance unit but also non-verbal information. For example, silence in face-to-face conversations can have a significant effect on the face, even for short periods of time.

Let $f(t)$ be the function on the two-dimensional plane with time (t) on the horizontal axis and $|De|$ on the vertical axis. At this time, the state where S and H are comfortable at a certain time point ($t = pt$) can be expressed by Eq. (3).

$$f(pt) < |\alpha| \tag{3}$$

The agent tries to make Eq. (3) hold continuously. If Eq. (3) does not hold, the agent verifies whether its behavior was impolite or too polite based on the sign of De. Also, the integral value (I) of De from the start time ($t = 0$) of the conversation between agent and user to a certain time point ($t = pt$) can be obtained by Eq. (4).

$$I = \int_0^{pt} f(t)dt \tag{4}$$

When the following equation is satisfied, the conversation between the agent and the user is comfortable at $t = pt$. That is, the agent tries to minimize I in conversation with the user.

$$I < \int_0^{pt} g(t)dt \ (g(t) = |\alpha|) \tag{5}$$

4 Questionnaire Study an Impact of Social Factors on Human Politeness Control

4.1 Method

(1) **Overview**
This study uses Google Forms as a survey platform. We refer to the politeness utterance evaluation method used in [5]. The relationship between the experimental participants and any person (S) is presented to the experimental participants. The participants subjectively evaluated "polite/impolite" (1: very impolite, 7: very polite) and "comfortable/uncomfortable" (1: very uncomfortable, 7: very comfortable) for the utterance related to the request by S. That is, in this experiment, the participants are H.

(2) **Social factors**
A "*Pardon?*" and "*Lend me money.*" were adopted as specific scenes. A "*Pardon?*" is considered to be meaningful to be investigated because it is considered that it frequently occurs in daily conversations between people and that the conversational

Table 1. Social factors.

D	P	Rx
$> \theta$: First meeting	$S > H$: Senior/boss	$> \theta$: "Lend me money." (expensive)
$= \theta$: Several months to one year of acquaintance	$S = H$: Equal	$< \theta$: Lend me money." (low price)
$< \theta$: More than one year of acquaintance	$S < H$: Junior/subordinate	

Table 2. Situations and politeness utterances.

Degree of politeness	*Pardon?*	*Lend me money*
6	*Mouichido itte-**morattemoyorosiidesyouka**?*	*Okarishite-**moyorosiidesyouka**?*
5	*Mouichido itte-**moraemasenka**?*	*Kasite-**moraemasenka**?*
4	*Mouichido itte-**kudasai***	*Kasite-**kudasai***
3	*Mouichido itte-**moratteii**?*	*Karite-**ii**?*
2	*Mouichido itte-**kureru**?*	*Kasite-**kureru**?*
1	*Mouichido itte* (WP: Without politeness)	*Kasite* (WP: Without politeness)

agent often conducts it to users. A "*Lend me money.*" is used only for communication between people, but was adopted because the *Rx* factor can be easily manipulated according to the amount borrowed. For the *D*, *P*, and *Rx* factors, we set each stage (Table 1). Hereafter, θ is a threshold for determining the magnitude of each factor.

(3) **Politeness utterances**
Utterances were created with reference to [9] in Japanese. Table 2 summarizes the utterance contents and the degree of politeness. The participants subjectively evaluate the utterances in Table 2 for each social factor.

(4) **Participants**
Experimental participants were gathered with the famous crowdsourcing service Cloud works. The total number of participants was 100 (44 females, average age = 39.7, *SD* = 9.7). All participants were Japanese native speakers.

4.2 Results

Firstly, we describe the result of "Pardon?". The politeness value generally agreed with the result of [9] (Fig. 1). On the other hand, SD is large for "moratteii?", "kur*eru?*" and WP. This is probably because the utterance with low politeness gave different impressions depending on the *D* and *P* factors. Also, the difference in comfortableness value between utterances was similar to that in politeness value (Fig. 2).

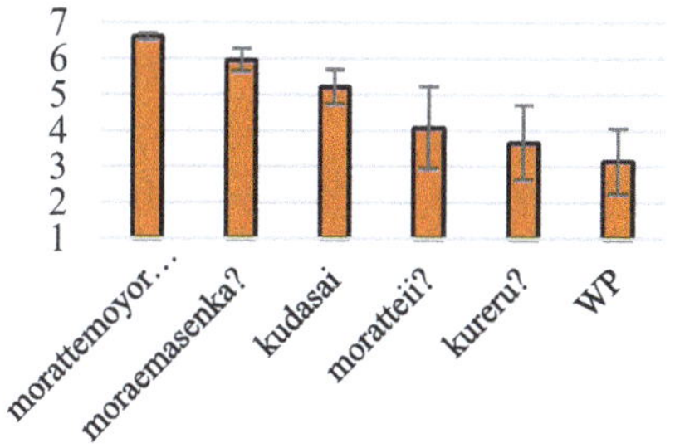

Fig. 1. Politeness value of "*Pardon?*".

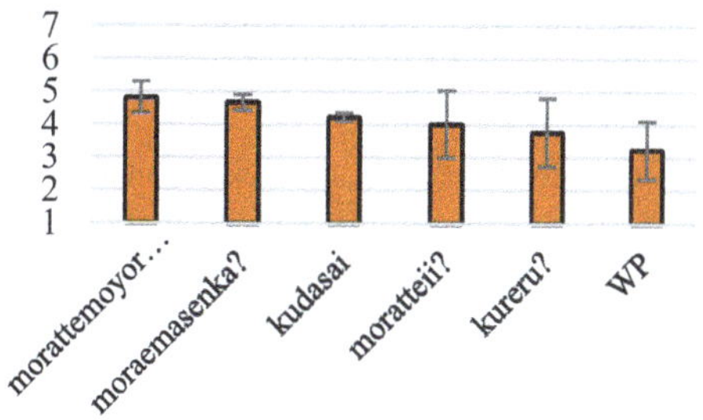

Fig. 2. Comfortableness value of "*Pardon?*".

We have performed a multiple regression analysis of the collected data set. The Eq. (6) was obtained for the impression value (*If*(*D*, *P*, *F*)) regarding politeness/impoliteness (coefficient of determination $R^2 = 0.583$).

$$If(D, P, F) - 0.27D + 0.64P + 0.74F + 1.39 \tag{6}$$

In Eq. (6), the*F* is the formalness of the utterance itself (degree of politeness). Also, the *If* is the impression of politeness (formal) and the larger the *If*, the higher the polite impression. Based on the regression equation, the farther the social distance (*D*) is, the less the polite impression, the *S*'s social power (*P*) is high, it is easy to feel polite. And the higher the politeness (*F*) of the utterance itself, the easier it is to feel polite.

The Eq. (7) was obtained for the impression value (*Ic*(*D*, *P*, *F*)) regarding comfortable/uncomfortable ($R^2 = 0.218$). *Ic* is the impression of comfortable. The larger the *Ic*, the higher the comfortable impression.

$$Ic(D, P, F) - 0.13D + 0.40P + 0.34F + 2.73 \tag{7}$$

Equation (7) shows that the farther the social distance (*D*) is, the less the comfortable impression. If *S*'s position (*P*) is high, it is easy to feel comfortable. Also, the higher the politeness (*F*) of the utterance itself, the easier it is to feel comfortable.

Similarly, multiple regression analysis was also applied to the "*Lend me money.*" ratings. The Eq. (8) was obtained for the impression value (*If*(*D*, *P*, *F*)) regarding politeness/impoliteness ($R^2 = 0.492$).

$$If(D, P, Rx, F) - 0.36D + 0.39P - 0.74Rx + 0.71F + 2.37 \tag{8}$$

Thus, the farther the social distance (*D*) is, the less the polite impression. If *S*'s position (*P*) is high, it is easy to feel polite. Also, the higher the *Rx*, the smaller the polite impression, the higher the politeness (*F*) of the utterance itself, and the easier it is to feel polite.

On the evaluation of comfortable, the Eq. (9) ($R^2 = 0.348$) was obtained for the impression value (*Ic*(*D*, *P*, *Rx*, *F*)).

$$Ic(D, P, Rx, F) - 0.32D + 0.28P - 0.56Rx + 0.50F + 2.68 \tag{9}$$

According to Eq. (9), the farther the *D* is, the less the comfortable impression, the *P* is high, it is easy to feel comfortable, the higher the *Rx*, the smaller the comfortable impression. And the higher the *F*, the easier it is to feel comfortable. The results of multiple regression analysis are summarized below.

- The farther the *D* is, the less polite the impression and the comfortable impression.
- The *S*'s *P* is high, it is easy to feel polite and comfortable.
- The higher the *Rx*, the smaller the comfortable impression.
- The higher the *F*, the easier it is to feel polite and comfortable.

5 Discussion

According to the experimental results, the regression model for evaluating politeness was reasonable. The results contribute as a model to select appropriate politeness of utterance according to *D*, *P*, *Rx* with the user in the proposed method. In particular, among the situations we employed in our experiments, the "*Pardon?*" occurs frequently in the interaction between the user and the conversational agent (e.g., smart speaker). Therefore, the usefulness of the regression model obtained in this experiment is high. Although "*Lend me money.*" is a situation that occurs only in conversations between people, the results of the experiments provide a valid regression model for the evaluation of politeness that takes the *Rx* factor into account. This suggests the importance of politeness control considering the *Rx* factor in our proposed method.

However, the coefficient of determination for comfort is low. Accurate estimation of the comfortable value is important for the facilitation of conversation. But, compared to politeness value, it is more difficult to estimate it on the basis of linguistic information only. For example, the results of the two-way ANOVA also showed that utterances with low politeness (e.g., "*moratteii?*") tended to give a significantly more comfortable impression to participants who had close relationships (*D* is low). Thus, in order to obtain a model that predicts the rating of comfortable, a complex set of factors must be taken into account. Specifically, top-down and discrete (and only three levels) *D* and *P* may not be enough to estimate comfortable with high accuracy. It may be necessary to use variables such as user attributes (gender, age, personality, etc.) and the *D* and *P* felt by the participants from the bottom-up.

6 Conclusion

In this paper, we proposed a method in which a conversational agent autonomously controls linguistic consideration (politeness). As a first approach, we examined how politeness is controlled in human conversation from the viewpoint of the impression for the experimental participants. The result of the experiment, a regression model was obtained with a coefficient of determination (R^2) of 0.58 for the impression of politeness, and a R^2 of 0.38 for the impression of comfortable. As future work, we will implement the politeness model in the conversational agent by referring to the obtained regression model and verify the usefulness of the model by interaction experiments. Our goal is to construct a conversational agent that chooses a conversational strategy for deepens the relationship as actively as possible within an error tolerance ($f(pt) < |\alpha|$).

Acknowledgments. This work was supported by JSPS KAKENHI Grant Number JP 18H03581 and 20H05572.

References

1. Brown, P., Levinson, S.C.: Politeness: Some Universals in Language Usage. Cambridge University Press, Cambridge (1987)
2. Usami, M.: Discourse politeness in Japanese conversation: some implications for a universal theory of politeness, Hituzi Syobo (2002)
3. Miyamoto, T., Katagami, D., Shigemitsu, Y.: Improving relationships based on positive politeness between humans and life-like agents. In: Proceedings of the HAI 2017, pp. 451–455 (2017)
4. Osawa, H., Yamada, S.: Social modification using implementation of partial agency toward objects. Artif. Life Robot. **16**(1), 78–81 (2011). https://doi.org/10.1007/s10015-011-0891-2
5. Danescu-Niculescu-Mizil, C., Sudhof, M., Jurafsky, D., Leskovec, J., Potts, C.: A computational approach to politeness with application to social factors. In: Proceedings of the ACL, pp. 250–259 (2013)
6. Zojaji, S., Peters, C., Pelachaud, C.: Influence of virtual agent politeness behaviors on how users join small conversational groups. In: Proceedings of the IVA 2020, Article No.: 59, pp. 1–8 (2020)
7. Face API. https://azure.microsoft.com/ja-jp/services/cognitive-services/face/. Accessed 30 Nov 2020
8. ST Emotion SDK. https://www.agi-web.co.jp/products/. Accessed 30 Nov 2020
9. Hill, B., Ide, S., Ikuta, S., Kawasaki, A., Ogino, T.: Universals of linguistic politeness: quantitative evidence from Japanese and American English. J. Pragmat. **10**(3), 347–371 (1986)

An Experimental Framework for Exploring the Social Dynamics of a Real-Time Physical Network Composed of Human Participants Using Beacon Technology

Zineb Elhamer(✉), Reiji Suzuki, and Takaya Arita

Nagoya University, Nagoya, Japan
zineb@alife.cs.i.nagoya-u.ac.jp

Abstract. We propose an experimental framework for understanding human social dynamics in a physically-situated and real-time decision-making environment. Participants have a Raspberry Pi Zero with an LCD, which uses Bluetooth to exchange signal strength and a game strategy between neighboring devices. They can move and change their strategy on the LCD in a closed space and asked to maximize the accumulated payoffs received from the neighbors, scaled by the signal strength, meaning that the distance between participants reflects their social closeness. We characterize the impact of information sharing among players on the overall social behavior by running preliminary experiments.

Keywords: Social dynamics · Cooperation · Contact networks · Physical networks · Social game · Social particle swarm

1 Introduction

Humans cooperate with each other within a network of social interactions whether virtual or real. Social networking services (SNSs) provide high social continuity as they keep us constantly connected in real-time in a non-discrete way, as well as in parallel through our handheld devices, as numerous people interact while connected at the same time. Additionally, SNSs comprise different topologies of interactions and hold communications at different frequencies, which presents continuous variations in the degree of social closeness with others. We have been investigating these continuous social dynamics using a social particle swarm (SPS) model[1] [3,6], in which particles move in a two-dimensional social or psychological space according to their social interactions among neighbors based on the particles' game-theoretical states (e.g., prisoner's dilemma) and positions, showing repeated emergence and collapse of cooperative clusters

[1] Link to video demos: https://vimeo.com/338902405.

K. Yada et al. (Eds.): JSAI 2020, AISC 1357, pp. 232–238, 2021.
https://doi.org/10.1007/978-3-030-73113-7_23

of individuals. These phenomena were observed in a web-based experimental framework for understanding dynamically changing social relationships in real-time decision-making and rewarding environments [8].

On the other hand, despite today's technological advancement which has reinforced the reliance on remote communication tools for interaction, several other studies have demonstrated that face-to-face communication is no less important than these tools [1,2,7].

Our purpose is to understand these emergent dynamics in a real-time social network composed of physically interacting participants in the context of the SPS model, using IoT[2]. In our experimental framework, participants are equipped with small devices and move around in a closed space, and the devices exchange Bluetooth signals, the strength of which represents the degree of their social effects. Players receiving the signal accumulate the score according to the signal strength and the sender and receiver's game-theoretical strategies which are displayed on the screen and can be changed through a button press. We introduce the design and implementation of the whole system and preliminary results.

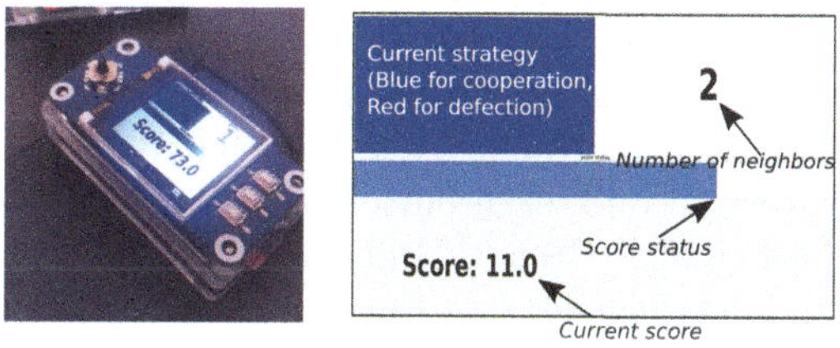

(I) (II)

Fig. 1. The interaction device and screenshot of the user interface.

2 The Proposed System

We designed and implemented a game-theoretical and experimental framework for understanding the social network of physically interacting human participants.

The experiment is carried out in an approximately $8.8 \times 8.45\,\text{m}^2$ space. Each player has a Raspberry Pi Zero WH device with an LCD hat mounted on it (Fig. 1(I)) which displays the necessary information for the player to ensure her proper integration within the game (Fig. 1(II)). Each participant has a strategy for the Prisoner's dilemma game (cooperate or defect), which is represented by the colored box on the top left of the screen. The screen also displays the score of the player accumulated over the course of each session at the bottom, and the amount of the received score at the current time represented by the blue status bar, where longer bars indicate a higher received score.

[2] We adopt EddyStone technology [4], which is a standard for beacons that operate on the Bluetooth Low Energy, and which is used for example by Google in The Physical Web project [5].

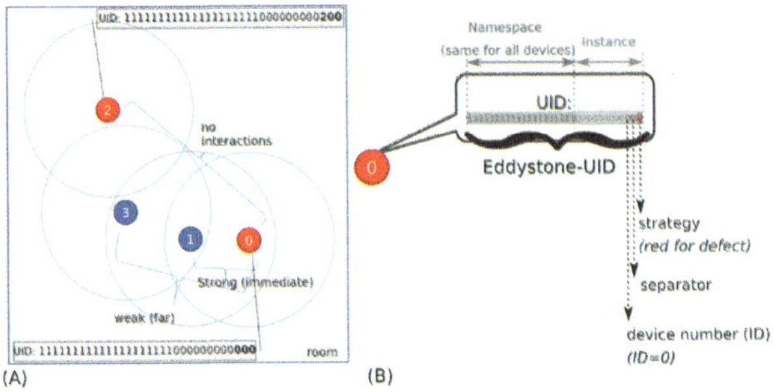

Fig. 2. An illustration of the flow of the experiment between each device, and the mechanism of neighbors' detection as well as information sharing between the devices.

Figure 2(A) represents a schematic illustration of participants in an experiment and information sharing between the devices based on the Bluetooth signals. The players interact according to the following payoff matrix values: $\{T, R, S, P\} = \{1.8, 1.0, -1.4, -1.0\}$, where the obtained score from the interaction with nearby players is calculated by summing up each individual payoff received from these players, then scaled by a social influence factor.

The social influence is represented by the signal strength value received from each device. Every second, each device measures the radio signal strength called RSSI (Receiver Signal Strength Indicator) when receiving packets from her neighbors. This value decreases as the distance between two given devices gets wide. Therefore, we can approximate the social closeness between the players by the strength of the received radio signal. We divided the covered zones into three categories based on the strength of the received signal which represents how socially close the player is, and discretized them into social closeness factors (Table 1).

Table 1. Estimation of the social closeness based on the received radio signal strength.

Signal interval	Signal category	Neighbor proximity	Social closeness value (s)
> -60	*Strong*	*Immediate*	1
$[-75, -60]$	*Medium*	*Near*	$\frac{1}{2}$
< -75	*Weak*	*Far*	$\frac{1}{3}$
No signal	No effect	Non-neighbor	N/A

At each time-step, the score is computed as follows:

$$score_i = \sum_{j \in neigh_i} pay_{(i,j)} \times s, \tag{1}$$

where s is a social closeness factor ($\in [0, 1]$) that translates how socially close two given players are, and $pay_{(i,j)}$ is the payoff i receives from neighbor j. $neigh_i$

is a set of other players of which signal value was received by i. The score is accumulated over the session as the total score.

As for the detailed implementation, we used DietPi operating system, which is lightweight and minimalistic and serves well our purpose. Inside each device, there is a program for conducting experiments with an interface that runs on startup. During each session, the program runs three threads in parallel; The first thread advertises the EddyStone UID of the beacon—composed of the ID of the device and its current strategy (cooperate or defect)—to other devices (Fig. 2 (B)). The second thread receives the Bluetooth signals from detected devices and extracts their IDs and strategies in order to calculate the received score. The last thread controls keypress events and flips the strategy color on the screen and updates the EddyStone UID.

3 The Preliminary Results

On February 21st 2020, we recruited five voluntary students from Nagoya University who are acquainted with the Prisoner's Dilemma game and conducted preliminary tests in the lecture room of the graduate school of informatics building at Nagoya University. The preliminary tests lasted for about 20 min. We asked the participants to maximize their own accumulated scores rather than their relative scores against others while roaming freely in the experiment space. The players hid their screens from each-other during the first three sessions, and in the last session, we asked them to make their screens visible to others in order to share their strategies more easily.

Four sessions were run with a short break in-between. Each session lasted for 5 min. In the results reported in this paper, we focus on (socially) close neighbors (i.e., the players recognized with the strong or medium social closeness), because they tended to be basically within the range of weak closeness because of the relative size of the room.

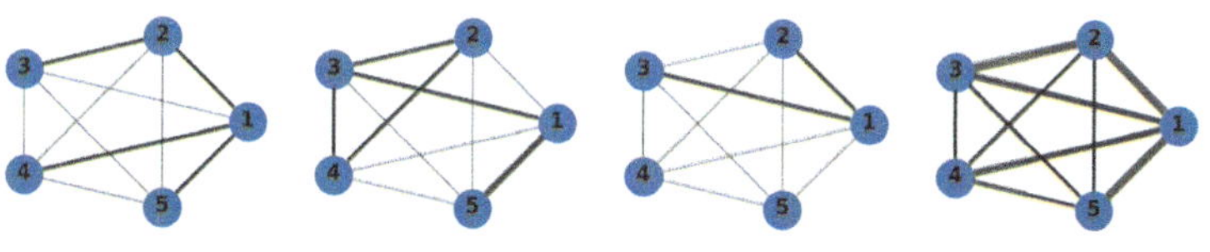

Fig. 3. A visualization of the resulting network of interactions in each session among close neighbors.

Figure 3 illustrates the layout of the network of interactions between the players in each session, where the nodes represent the players and the links between the nodes represent the interactions that happen between two corresponding players, where the thickness of the link indicates the frequency of interactions (i.e., the count of being recognized as neighbors with each other at the same timing) between two nodes. The thicker the line, the more frequent interactions have occurred between the corresponding players. We see that in the last session, where information sharing was permissible, interactions between players were more frequent than in the previous sessions.

We also extracted the data from each device during each session and generated a dynamic visualization of the temporal network, which can be viewed on the following link: https://vimeo.com/458608711. In this animation, each timestep, a highlighted edge represents an active interaction that is happening within the strong (black) and medium (gray) social closeness. We see that in the last session where the strategy information was shared, more active edges with strong or medium closeness were formed which sustains our previous observation.

We focus on the number of participants with strong or medium social closeness (which are called (socially) close neighbors hereafter). Figure 4(A) represents the temporal distribution of the number of neighbors of each player during the four sessions, which is shown in black dots in the first five columns (representing the overall number of players). We see that during the first three sessions where sharing the strategy information did not happen, this number tended to be zero most of the time. By contrast, the results in the last session show a wide variation in both the number of neighbors and the strategy update rate which was reactively changing for each player. This is because players were tempted to get closer to each other to obtain the information about the others, and this also made them interactive and resulted in mutual cooperation.

On the same figure, the last two columns represent the distribution of the accumulated scores and the proportion of cooperators, with brighter colors denoting higher values. The overall proportion of cooperators throughout the experiment was dynamically changing in all sessions but was comparatively higher when information sharing was allowed (session 4). This is because they were actively engaged in interactions to establish cooperative relationships, instead of being individually isolated which leads to fewer chances of accumulating more scores.

Figure 4(B) represents the change in the total score of each player as a function of time. We see that in the hidden information scenario (sessions 1 to 3), the total scores varied among players due to the absence of information about the strategies of the other players, which made them less interactive, and often lose chances to form mutual relationships and the means to decide a better strategy given the circumstance. In session 4, the total scores of the players were constantly ascending simultaneously and reached high values. This is due to the additional instruction of sharing strategy information, which would suggest the importance of information sharing for successful cooperation.

We further compare two cases of no-information versus with-information sharing, Fig. 5 represents the adjacency matrices of the network of players taken from the first and last sessions, juxtaposed next to each other, where the blue color represents a cooperation strategy, and red represents defection. We see that in the last trial, more interactions were spotted, and the strategies of the players were mostly cooperative. Another observation is that cooperative strategies hap-

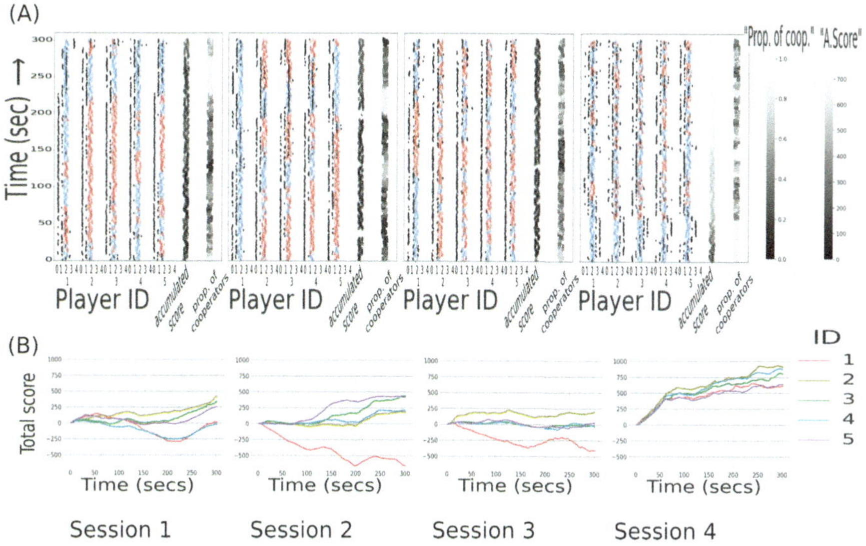

Fig. 4. (A): In each of four sub-figures, the left four columns represent the temporal changes (from bottom to top) in the players' strategies, as well as the number of their close neighbors (in black dots), the right two columns, represent the distribution of the accumulated scores and the proportion of cooperators in the network respectively, with brighter colors denoting higher values. (B): The temporal change in the total scores of each player.

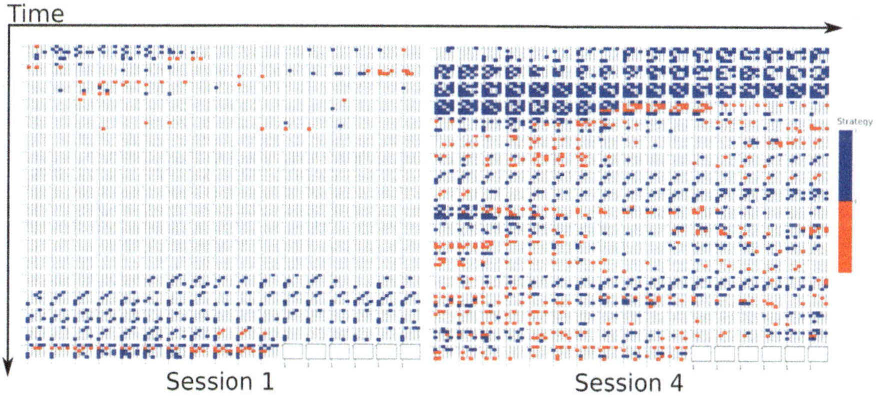

Fig. 5. The juxtaposition of the adjacency matrices with the strategy color representations (red for defection; blue for cooperation) at close social proximity from session 01 and 04 over the course of the experiment. The wide x and y axes represent the flow of time, starting from left to right, top to bottom. Each square contains the 5 × 5 adjacency matrix of the players at a moment t along with their strategy colors. The x and y-axis in each square represent a pair of players. The blue and red dots represent that the x recognized y as a close neighbor and cooperated and defected against y, respectively.

pen collectively, that is when someone chooses to cooperate, she is surrounded by a majority of cooperators, whereas defection strategies happen individually, where she doesn't find a neighbor willing to interact.

4 Conclusion

We designed and implemented a game-theoretical and experimental framework for understanding the social network of physically interacting human participants. Information visibility among players allowed them to adapt to each other's strategies by actively switching their screen colors. It also helped boost cooperation in the physical network through a chain reaction effect where a small number of cooperators could trigger global cooperation within the network. However, this behavioral pattern soon loses its consistency and collapses due to the emergence of some defectors, this sea-saw effect was also observed in the SPS model via the formation and collapse of cooperative assortments.

Acknowledgements. This work was supported in part by JSPS/MEXT KAKENHI JP17KT0001 and Topic-Setting Program to Advance Cutting-Edge Humanities and Social Sciences Research JP17J0011b.

References

1. Nardi, B., Whittaker, S.: The place of face-to-face communication in distributed work. In: Hinds, P., Kiesler, S. (eds.) Distributed Work, pp. 83–111. MIT Press, Cambridge (2002). https://doi.org/10.7551/mitpress/2464.003.0008
2. Eagle, N., Pentland, A.: Reality mining: sensing complex social systems. Pers. Ubiquit. Comput. **10**, 255–268 (2006). https://doi.org/10.1007/s00779-005-0046-3
3. Elhamer, Z., Suzuki, R., Arita, T.: The effects of population size and information update rates on the emergent patterns of cooperative clusters in a large-scale social particle swarm model. Artif. Life Robot. **25**(1), 149–158 (2019). https://doi.org/10.1007/s10015-019-00558-6
4. Google: Eddystone protocol specification. https://github.com/google/eddystone
5. Google: The Physical Web (2014). https://google.github.io/physical-web
6. Nishimoto, K., Suzuki, R., Arita, T.: Social particle swarm: explosive particle dynamics based on cooperative/defective forces. In: Proceedings of 2013 IEEE Symposium on Artificial Life (IEEEALIFE2013), pp. 134–139 (2013). https://doi.org/10.1109/alife.2013.6602442
7. Storper, M.: Buzz: face-to-face contact and the urban economy. J. Econ. Geogr. **4**(4), 351–370 (2004). https://doi.org/10.1093/jeg/4.4.351
8. Suzuki, R., Ito, M., Kodera, S., Nishimoto, K., Arita, T.: An online experimental framework for cooperative relationships with a real-time decision-making and rewarding environment. Front. Ecol. Evol. **6**(74) (2018). https://doi.org/10.3389/fevo.2018.00074

New Approach to Estimating Mental Health Score Using a Communication Agent

Satoshi Nakagawa(✉), Shogo Yonekura, Hoshinori Kanazawa, Satoshi Nishikawa, and Yasuo Kuniyoshi

The University of Tokyo, 7-3-1 Hongo, Bunkyo-ku, Tokyo, Japan
nakagawa@isi.imi.i.u-tokyo.ac.jp
http://www.isi.imi.i.u-tokyo.ac.jp

Abstract. It is important for monitoring systems for the elderly to estimate the state of the user during an interaction. In the field of elderly welfare, quality of life (QOL) is a useful indicator for comprehensive treatment of physical suffering as well as mental health and social well-being. In this study, we propose a new approach to QOL estimation that integrates facial expressions, head fluctuations, and eye movements captured during interactions with a communication agent. To this end, we implemented a communication agent and constructed a database based on the information collected through interpersonal experiments with 14 participants. In addition, we implemented a multimodal learning estimator and performed training with facial expressions, head fluctuation, and gaze feature extraction. Our results show that integration learning realized with all of the features integrated provided less error than single-modal learning based on facial expression alone. From our experimental results, we concluded that the proposed system can be used adequately as a QOL estimator.

Keywords: Quality of life · Human-agent interaction · Elderly welfare

1 Introduction

The global population of people 65 years and older was about 703 million in 2019. This number is expected to increase to 1.5 billion by 2050 [21]. Therefore, in recent years, monitoring systems have been implemented in the daily lives of the elderly to ensure their welfare. When such monitoring system interacts with a human, it is important for the system to accurately estimate their state. In the field of elderly welfare, quality of life (QOL) is a useful indicator [1] that can be used to determine not only human physical condition but also mental health and social well-being.

In this study, we refer to the QOL index of the SF-36v2 health survey [7]. According to this index, QOL consists of eight subscales: Physical Functioning

K. Yada et al. (Eds.): JSAI 2020, AISC 1357, pp. 239–247, 2021.
https://doi.org/10.1007/978-3-030-73113-7_24

(PF), Role Physical (RP), Body Pain (BP), General Health (GH), Vitality (VT), Social Functioning (SF), Role emotional (RE), and Mental Health (MH). The questionnaire provided in the SF-36v2 consists of 36 questions; each question is answered by choosing one of three to six choices. The answers are fed into a norm-based scoring algorithm with standardized values for the national population, with a score calculated for each scale (Fig. 1 left).

The user's QOL can be evaluated through the conventional measurement method using a questionnaire. However, this approach is problematic because it requires effort from the user and cannot be used to immediately track changes in QOL scores. Therefore, we propose a system that learns the relationship between each QOL scale and information obtained from the user through a communication agent and estimates the QOL scores based on that information (Fig. 1 right).

To this end, we implemented a communication agent and constructed a database based on the information collected through communication experiments with humans. A previous study [14] proposed QOL score estimation from facial expressions, but the Mental Health QOL scale score was found to be difficult to estimate. Therefore, in this study, we propose a new approach to QOL estimation that integrates facial expressions, head movements, and eye movements, which are important features in emotion recognition [11,25]. Furthermore, these features can be extracted as visual information in the process of interacting with a communication agent.

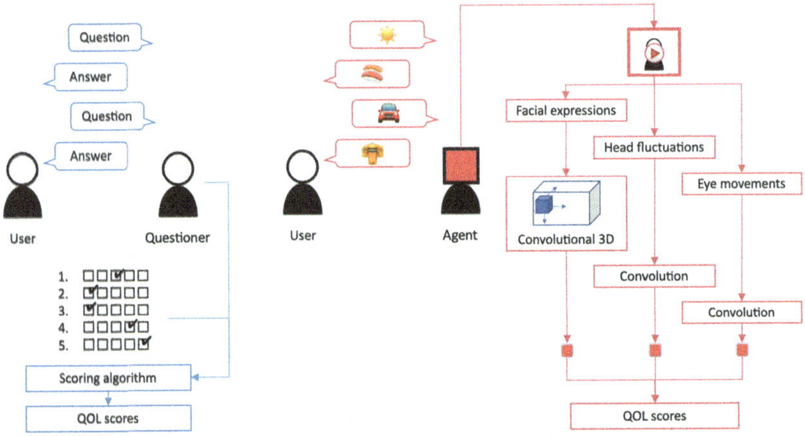

Fig. 1. Quality of life (QOL) estimation during interaction with an agent

2 Related Works

There has been a long history of research activities to establish a method for estimating the current state of a person. These approaches include vision-based, audio-based, text-based, electroencephalogram (EEG)-based, sensor-based, and multimodal approaches. There are many fields where these approaches and systems can be applied, but with the aging society, it is becoming more important to devise a monitoring system for the elderly. In particular, it is important to develop a non-invasive system that does not burden the users. Solbach and Tsotsos developed a fall detection system for elderly people using OpenPose [2], a two-dimensional posture estimation method [19]. In addition, Mori et al. developed an abnormality detection system based on the estimation of elderly people's living patterns [13]. In the area of mental health, Ren et al. proposed a system for estimating trends in mental states based on tweets on Twitter [18] and based on textual data from daily blogs, which were applied to predicting suicidal behavior [17]. Higuchi et al. also proposed a method that detected the presence of mental health disorders from speech information [8]. Furthermore, DeVault et al. implemented an interactive agent in a virtual space for interactive state estimation and appropriate response generation [4].

As for current state estimation, QOL has been attracting attention as an indicator for the pluralistic understanding of the state of the elderly. However, most studies treating QOL considered the impact of their proposed method on QOL scores, and estimation of QOL itself in the human-robot interaction (HRI) field appeared for the first time in 2018 [14]. That study, in which a prediction model was trained using videos as input, showed that the Mental Health QOL scale was the most difficult scale to estimate. The reported reasons for this were that information on mental health was missing from the training data or that although the video data contained information on mental health, the system failed to extract it adequately. Therefore, Nakagawa et al. proposed an improved estimation method based on gaze patterns and suggested that it may be useful for estimating mental health scores [15].

Currently, improvements in eye-tracking techniques and the ability to use facial features for estimation have led to an increasing number of studies on estimation methods based on gaze patterns. Several studies [12,25] have designed emotion recognition systems based on gaze patterns in combination with EEG signals, which objectively and accurately reflect an individual's emotional state [16]. Similarly, studies have been conducted to achieve estimation of emotional state by integrating multiple features, such as data extracted from human faces. Hu et al. achieved facial expression recognition by integrating multiple feature maps and facial patches, such as eyebrows, eyes, and mouth [9]. Furthermore, Zeng et al. reported that expression estimation studies have been conducted using head movements in combination with feature points related to facial expressions [24], eyebrow motion [3], and facial surface temperature [11,23].

3 Communication Agent and Interpersonal Experiment

We implemented a free communication agent on a personal computer and constructed a database of information collected through interpersonal experiments. The communication agent was implemented using a chat API developed by NTT

Docomo [5]. It consisted of four parts: voice input, voice recognition, reply generation, and voice synthesis as output. The agent was able discuss a wide range of topics. After the user completed the SF-36v2 QOL questionnaire and the QOL scores were calculated using the existing scoring algorithm, the user held a free conversation with the communication agent. During this conversation, we collected videos of the user's upper body (Fig. 2). The experiment involved 14 adults (7 males and 7 females). All participants received prior explanation of the experiment and freely provided confirmation of their consent. During the experiment, no restrictions were made regarding the content of the conversation or total conversation time. Through the above experiments, we collected video and audio data, QOL questionnaire results, and actual QOL scores.

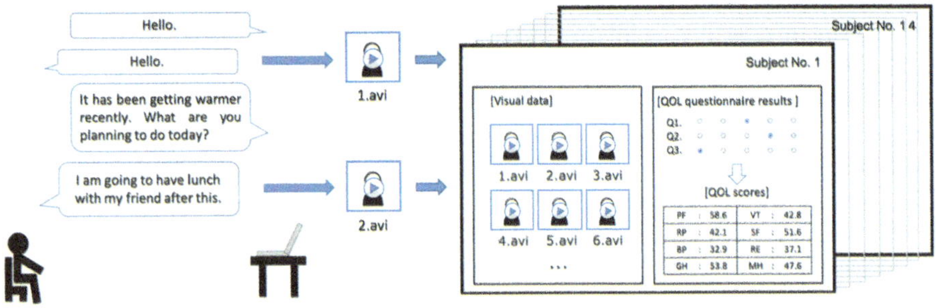

Fig. 2. Interpersonal experiment using conversational agent

4 Multiple Feature Vector Learning System

To obtain a multimodal learning estimator, we improved the learning estimation architecture of the QOL estimation system proposed in [14]: we implemented a system that integrates the output of each estimator at the final stage using facial expression, head fluctuation, and gaze features individually (Fig. 1 right). For facial expression extraction, frame division was performed with a frame rate of 20, and a face detector called Haar Cascades [22] was applied to each frame. Regarding head fluctuation and gaze feature extraction, facial feature points were extracted using OpenPose [2], and time-series data was extracted from their positions. We applied OpenPose to the extraction of gaze patterns and head movements to unify the method for extraction of both features, so that information obtained from the camera on the robot could be used. In the learning estimation stage, Convolutional 3D (C3D) [20], an approach that performs convolution combining two spatial dimensions and one temporal dimension, was applied for learning using facial expressions. This C3D-based network has been shown to be effective in video-based emotion recognition [6]. Convolutional 1D was applied for each learning process using head fluctuation and gaze pattern. In the final stage, we performed the QOL score regression with each estimation result being used as input to estimate the QOL score directly. We applied the leave-one-out cross-validation method [10] to verify the proposed model.

5 Experimental Results

To verify the effectiveness of the proposed model, we focused on the score estimation accuracy of the Mental Health QOL scale. Figure 3 shows the distribution of error between the estimated and actual QOL scores in the proposed multimodal learning and in single-modal learning using facial expressions only. It was shown that multimodal learning was able to perform QOL score estimation with smaller estimation error than the single-modal learning (multimodal: 5.6 ± 4.6 vs. single-modal: $14.6 \pm 10.2, p < .05$). Furthermore, the results of the Mann-Whitney U test performed between integrated and single-modal learning show that the estimation results using the proposed method succeeded in significantly reducing the error compared to the conventional method.

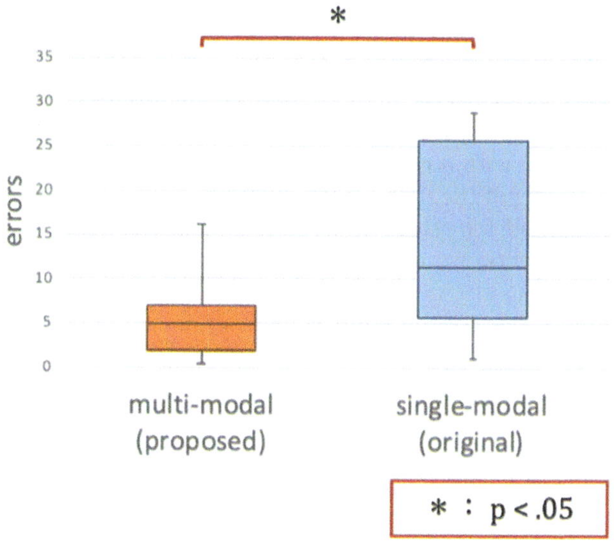

Fig. 3. Distribution of errors

6 Discussion

In this study, a robust estimator of Mental Health QOL states was created from extracted features of head fluctuations and gaze patterns, and by integrating them with facial expressions. This estimator succeeded in extracting features related to Mental Health QOL states from the visual information. The results obtained by integrating the three features were significantly more accurate than single-modal learning using facial expressions. At the same time, the superiority of multimodal over single-modal method is demonstrated by the results. In addition, the QOL score derived by the conventional method is adjusted so that the average is 50 points and the standard deviation is 10. Therefore, we concluded that the proposed method can be utilized as a mental health estimator because the mean of the estimation error was less than 10.

In this study, we compared estimation accuracy under two conditions: integrated learning and single-modal learning, but, in both conditions, the databases we used were built through conversation experiments with our original implementation of interactive agents. However, it was difficult to estimate the single-modal learning condition of the previous study [14], even though the same database was used. There are two reasons for this. First, information about the head fluctuation was missing from the input data when the faces were extracted in the preprocessing for input to the training model C3D. The second reason is that the amount of change other than the gaze pattern in the videos of facial expressions was large, and the gaze pattern was not properly extracted and learned. The estimator based on integration learning proposed in this experiment solved these problems and thus improved the estimation accuracy.

In this study, we proposed a comprehensive state estimation method using QOL as an indicator, which is an alternative to emotion estimation, the mainstream method of state estimation today [4,16]. We estimated the user's state during an interaction with a conversational agent and realized a broad and pluralistic understanding of the user's state, which was not limited to emotions. By establishing a method for estimating the variability of QOL of elderly people using interactive agents, we can present a new approach for estimating their current states.

One of the limitations of this study is its focus on the early stages of relationship building between a human and an agent. The relationship between humans and robots changes with the passage of time, and the information the robot acquires from the user is expected to change as well. In order for a robot to continue to estimate the state of the human over time, it is necessary to construct an estimation model based on long-term data. Therefore, as a future direction, we will create an estimator based on data from long-term interactions that also reflects changes in the interactions over time. The implementation of this estimator will improve the quality of interactions and answer one of the main questions of human-robot interaction (HRI): how to establish a long-term and continuous relationship between humans and robots.

7 Conclusion

In this study, we proposed a Mental Health QOL score estimator, a new approach of user understanding, by integrating features extracted from time series data collected from participants during their interactions with a communication agent. The experimental results show the effectiveness of the proposed multimodal learning model in the direct estimation of the Mental Health QOL scale, with estimation error of less than 10. Thus, we concluded that the proposed system is sufficiently accurate to be used as a QOL estimator. Furthermore, the superiority of multimodal learning over single-modal learning was proved.

For future research, we plan to further improve the estimation function by integrating and learning all the information obtained in the interaction process, such as prosodic and text information. We will begin research on human-agent interaction based on QOL estimation for individual-specific monitoring. Moreover, we will work on the creation of robot utterances and behaviors specialized for the user, based on their current state and determined by our multimodal QOL estimation system.

Acknowledgements. This research was supported by the KDDI Foundation Research Grant Program 2019 and Graduate Program for Social ICT Global Creative Leaders (GCL) of The University of Tokyo by MEXT (Ministry of Education, Culture, Sports, Science and Technology).

References

1. Cabinet Office (CAO): Annual report on the ageing society 2018 (2018). https://www8.cao.go.jp/kourei/english/annualreport/2018/2018pdf_e.html
2. Cao, Z., Simon, T., Wei, S.-E., Sheikh, Y.: Realtime multiperson 2D pose estimation using part affinity fields. In: IEEE Conference on Computer Vision and Pattern Recognition (CVPR), pp. 1302–1310 (2017). https://doi.org/10.1109/CVPR.2017.143
3. Cohn, J.F., Reed, L.I., Ambadar, Z. Xiao, J., Moriyama, T.: Automatic analysis and recognition of brow actions and head motion in spontaneous facial behavior. In: 2004 IEEE International Conference on Systems, Man and Cybernetics (IEEE Cat. No. 04CH37583), vol. 1, pp. 610–616. IEEE (2004). https://doi.org/10.1109/ICSMC.2004.1398367
4. DeVault, D., et al.: SimsenSei Kiosk: a virtual human interviewer for healthcare decision support. In: Proceedings of the 2014 international Conference on Autonomous Agents and Multi-agent Systems, pp. 1061–1068. International Foundation for Autonomous Agents and Multiagent Systems (2014)
5. docomo developer support, NTT Docomo. (n.d.). https://dev.smt.docomo.ne.jp. Accessed 26 Feb 2020
6. Fan, Y., Lu, X., Li, D., Liu, Y.: Video-based emotion recognition using CNN-RNN and C3D hybrid networks. In: Proceedings of the 18th ACM International Conference on Multimodal Interaction, pp. 445–450 (2016). https://doi.org/10.1145/2993148.2997632
7. Fukuhara, S., Bito, S., Green, J., Hsiao, A., Kurokawa, K.: Translation, adaptation, and validation of the SF-36 health survey for use in Japan. J. Clin. Epidemiol. **51**(11), 1037–1044 (1998). https://doi.org/10.1016/S0895-4356(98)00095-X
8. Higuchi, M., et al.: Discrimination of bipolar disorders using voice. In: Cipresso, P., Serino, S., Villani, D. (eds.) MindCare 2019. LNICSSITE, vol. 288, pp. 199–207. Springer, Cham (2019). https://doi.org/10.1007/978-3-030-25872-6_16
9. Hu, M., Yang, C., Zheng, Y., Wang, X., He, L., Ren, F.: Facial expression recognition based on fusion features of center-symmetric local signal magnitude pattern. IEEE Access **7**, 118435–118445 (2019)

10. Kocaguneli, E., Menzies, T.: Software effort models should be assessed via leave-one-out validation. J. Syst. Softw. **86**(7), 1879–1890 (2013). https://doi.org/10.1016/j.jss.2013.02.053
11. Liu, P., Yin, L.: Spontaneous facial expression analysis based on temperature changes and head motions. In: 2015 11th IEEE International Conference and Workshops on Automatic Face and Gesture Recognition (FG), vol. 1, pp. 1–6. IEEE (2015). https://doi.org/10.1109/FG.2015.7163094
12. Lu, Y., Zheng, W.L., Li, B., Lu, B.L.: Combining eye movements and EEG to enhance emotion recognition. In: Proceedings of the 24th International Conference on Artificial Intelligence (IJCAI), pp. 1170–1176. AAAI Press (2015)
13. Mori, T., et al.: Life pattern estimation of the elderly based on accumulated activity data and its application to anomaly detection. J. Robot. Mechatron. **24**(5), 754–765 (2012). https://doi.org/10.20965/jrm.2012.p0754
14. Nakagawa, S., Enomoto, D., Yonekura, S., Kanazawa, H., Kuniyoshi, Y.: A telecare system that estimates quality of life through communication. In: 2018 5th IEEE International Conference on Cloud Computing and Intelligence Systems (CCIS), pp. 325–330 (2018). https://doi.org/10.1109/CCIS.2018.8691360
15. Nakagawa, S., Enomoto, D., Yonekura, S., Kanazawa, H., Kuniyoshi, Y.: New telecare approach based on 3D convolutional neural network for estimating quality of life. Neurocomputing **397**, 464–476 (2020). https://doi.org/10.1016/j.neucom.2019.09.112
16. Ren, F., Dong, Y., Wang, W.: Emotion recognition based on physiological signals using brain asymmetry index and echo state network. Neural Comput. Appl. 1–11 (2018). https://doi.org/10.1007/s00521-018-3664-1
17. Ren, F., Kang, X., Quan, C.: Examining accumulated emotional traits in suicide blogs with an emotion topic model. IEEE J. Biomed. Health Inform. **20**(5), 1384–1396 (2015). https://doi.org/10.1109/JBHI.2015.2459683
18. Ren, F., Wu, Y.: Predicting user-topic opinions in Twitter with social and topical context. IEEE Trans. Affect. Comput. **4**(4), 412–424 (2013). https://doi.org/10.1109/T-AFFC.2013.22
19. Solbach, M.D., Tsotsos, J.K. Vision-based fallen person detection for the elderly. In: Proceedings of the IEEE International Conference on Computer Vision Workshops, pp. 1433–1442 (2017). https://doi.org/10.1109/ICCVW.2017.170
20. Tran, D., Bourdev, L., Fergus, R., Torresani, L., Paluri, M.: Learning spatiotemporal features with 3D convolutional networks. In: Proceedings of the IEEE International Conference on Computer Vision, pp. 4489–4497 (2015). https://doi.org/10.1109/ICCV.2015.510
21. The United Nations: World population ageing 2019 (2019). https://www.un.org/en/development/desa/population/publications/pdf/ageing/WorldPopulationAgeing2019-Report.pdf
22. Viola, P., Jones, M.: Rapid object detection using a boosted cascade of simple features. In: Proceedings of the 2001 IEEE Computer Society Conference on Computer Vision and Pattern Recognition, p. I (2001). https://doi.org/10.1109/CVPR.2001.990517
23. Zeng, Z., Pantic, M., Roisman, G.I., Huang, T.S.: A survey of affect recognition methods: audio, visual, and spontaneous expressions. IEEE Trans. Pattern Anal. Mach. Intell. **31**(1), 39–58 (2009). https://doi.org/10.1109/TPAMI.2008.52

24. Zhang, Y., Ji, Q.: Active and dynamic information fusion for facial expression understanding from image sequences. IEEE Trans. Pattern Anal. Mach. Intell. **27**(5), 699–714 (2005). https://doi.org/10.1109/TPAMI.2005.93
25. Zhao, L.M., Li, R., Zheng, W.L., Lu, B.L.: Classification of five emotions from EEG and eye movement signals: complementary representation properties. In: 2019 9th International IEEE/EMBS Conference on Neural Engineering (NER), pp. 611–614 (2019). https://doi.org/10.1109/NER.2019.8717055

Author Index

K. Yada et al. (Eds.): JSAI 2020, AISC 1357, pp. 249–250, 2021.
https://doi.org/10.1007/978-3-030-73113-7

The manufacturer's authorised representative in the EU is Springer Nature Customer Service Centre GmbH, Europaplatz 3, 69115 Heidelberg, Germany. If you have any concerns regarding our products, please contact ProductSafety@springernature.com

Printed and bound by CPI Group (UK) Ltd, Croydon, CR0 4YY
30/09/2026
02225251-0005